OF MEN AND SEALS

A HISTORY OF THE NEWFOUNDLAND SEAL HUNT

JAMES E. CANDOW

Studies in Archaeology
Architecture and History

National Historic Parks and Sites
Canadian Parks Service
Environment Canada

Available in Canada through authorized bookstore agents and other bookstores, or by mail from the Canadian Government Publishing Centre, Supply and Services Canada, Hull, Quebec, Canada K1A 0S9.

Published under the authority
of the Minister of the Environment,
Ottawa, 1989.

Editing and design:Jean Brathwaite
Index:J. Peterson

Parks publishes the results of its research in archaeology, architecture, and history. A list of publications is available from Research Publications, Canadian Parks Service, Environment Canada, 1600 Liverpool Court, Ottawa, Ontario K1A 0H3.

Canadian Cataloguing in Publication Data

Candow, James E.

Of men and seals: a history of the Newfoundland seal hunt.

(Studies in archaeology, architecture and history,
ISSN 0821-1027)
Issued also in French under title: Des hommes et des phoques.
Includes bibliography and index.
ISBN 0-660-12938-8
DSS cat. no. R61-2/9-41E

1. Sealing — Newfoundland — History. 2. Sealers (Persons) — Newfoundland — History. 3. Sealing — Canada — History. I. Canada. National Historic Parks and Sites. II. Title. III. Title: A history of the Newfoundland seal hunt. IV. Series.

SH362.C35 1988   C88-097034-0   639'.29'09718

60,831

Contents

Contents

Preface

This study began as an attempt to accumulate a research base for a Canadian Parks Service commemorative exhibit on the Newfoundland seal hunt. It soon became apparent that numerous works treated specific periods in the history of the hunt, but no one work covered them all. Although it is not definitive, this book attempts to fill that void. Because of time constraints, I could make little use of primary sources and only rarely was able to consult people who had participated in the hunt. However, it does touch on the more important themes and will serve as an introduction to the subject.

I have avoided the term "seal fishery." The fact that the seal lives in the sea and can be caught with nets has given rise to this usage, but it is nevertheless technically incorrect. The seal is a mammal. Of course, many object to the term "seal hunt," maintaining that the operation was a slaughter. But because sealers ventured out in search of their prey, the word "hunt" is legitimate. Finally, I have observed popular rather than scientific nomenclature in using "hood" as opposed to "hooded" seal.

I am grateful to numerous individuals and institutions for their assistance. Captain Morrissey Johnson, M.P., of St. John's, owner of Johnson Combined Enterprises; Mr. Harold Henriksen of Halifax, general manager of Karlsen Shipping Company Limited; and Captain Alfred M. Shaw of Halifax, former owner of Mayhaven Shipping, all shared with me their extensive knowledge of the hunt. Mr. Royal Cooper of Gander, Newfoundland, who served as a seal-spotter pilot with Maritime Central Airways and Eastern Provincial Airways from 1952 to 1975, was an invaluable source for the post-1947 aerial spotting service. Mr. H.M. Andrews of the Workers' Compensation Commission of Newfoundland and Labrador explained the intricacies of recent regulations governing workers' compensation for sealers. The staffs of the

Centre for Newfoundland Studies, Memorial University of Newfoundland, and the Provincial Archives of Newfoundland and Labrador offered their usual friendly and helpful service. I am also grateful to Dr. Shannon Ryan and Dr. J.K. Hiller of the Department of History, Memorial University of Newfoundland, for taking the time to read early drafts of the work. Dr. W.D. Bowen, chief of the Marine Fish Division, of the Bedford Institute of Oceanography supplied several photographs of the modern hunt and also corrected portions of the manuscript dealing with scientific questions.

Submitted for publication 1985 by James E. Candow, Atlantic Regional Office, Canadian Parks Service, Environment Canada, Halifax.

Introduction

Harps and Hoods

The Newfoundland seal hunt was based on the harp and, to a lesser extent, hood seal species.[1] There are other North Atlantic seals, such as the square flipper, ringed, bearded, and small Greenland seals, but they are not as numerous. The harp and hood are hair seals (family Phocidae) valued for their skin, fat, and hair.

The harp seal (*Phoca groenlandica*) has four main breeding areas: off northern Newfoundland and southeastern Labrador (the "Front"); the Gulf of St. Lawrence; the White Sea; and off the east coast of Greenland near Jan Mayen. Except when they moult, breed, and give birth (whelp), harp seals spend all their time in the water, pursuing their migratory lives. The summer feeding grounds of the western Atlantic, or Newfoundland, harps are located off Baffin Island, their extreme range extending farther north to Devon and Ellesmere islands. The seals begin their southward migration in September or October with the onset of winter ice and usually pass Cape Chidley, at the northern tip of Labrador, by early November. Towards the end of December the herd, now near the Strait of Belle Isle, splits into two sections, one-third going through the strait into the Gulf of St. Lawrence and the other two-thirds following on down the northeast coast of Newfoundland. Recent evidence indicates some slight intermingling between the two herds. The seals eventually scatter south off the Newfoundland coast, where an ample

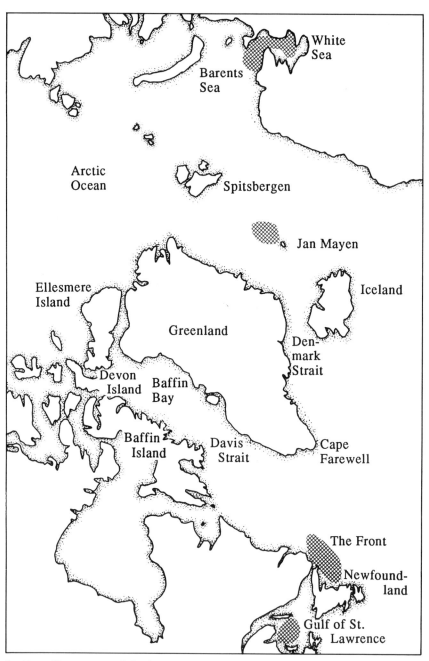

1 Breeding areas of the harp seal.
Map by Wayne Hughes.

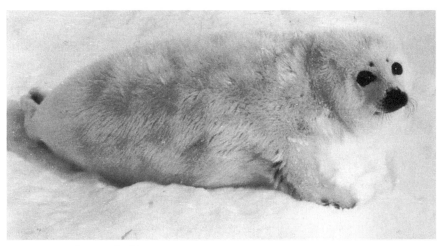

2 Harp seal pup ("whitecoat"), six days old.
Canada, Department of Fisheries and Oceans, St. John's; W.D. Bowen photo.

3 Adult harp seal.
Canada, Department of Fisheries and Oceans, St. John's; W.D. Bowen photo.

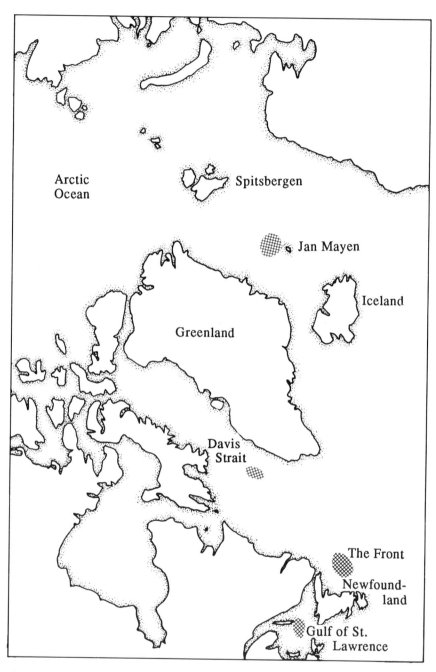

4   Breeding areas of the hood seal.
Map by Wayne Hughes.

5 Hood seal pup ("blueback").
Canada, Department of Fisheries and Oceans, St. John's; W.D. Bowen photo.

6 Adult male hood seal.
Canada, Department of Fisheries and Oceans, St. John's; W.D. Bowen photo.

supply of marine life sustains them. After heavy feeding from December to February, particularly by the females, the harps begin the return trip, emerging onto the ice to whelp in late February or early March off Newfoundland (about five days earlier in the gulf). In recent years the whelping grounds at the Front have tended to be located farther north.[2] The females give birth and, as soon as the pups are weaned, begin to mate. After mating, the adults briefly take to the water again only to return to the ice in April to moult. Once they have moulted, by early May, they resume the journey to their summer feeding grounds.

Harp seal pups formed the chief object of the seal hunt. Although it weighs only 20 pounds at birth, the pup grows rapidly, reaching 75 pounds within three weeks. Its growth is made possible by the female harp seal's rich milk, which contains over 40 per cent fat. Cow's milk, by comparison, is only 4 per cent fat. In terms of commercial value, the pup is at its peak six to ten days after birth, when its fat will yield the greatest amount of oil. The lanugo (the white fetal coat, hence "whitecoat") that became important in the 20th century is present during the first two weeks. Thereafter the pup begins to moult and coarser hair appears. Pups at this stage are known as "raggedy jackets." After weaning, 12 to 15 days after birth, the pup loses weight and its value decreases accordingly. The pup takes to the water for the first time around the end of March, feeding mainly on small crustacea. All pups start swimming north in May and June.

From the time they take to the water until they reach their first year, the young harps are called "beaters." From then until they attain breeding age they are "bedlamers," after the French *bêtes de la mer*. Harps of both sexes usually mature sexually by their fourth year. They then possess the distinctive markings from which they derive their name — a light grey coat with a band of black spots beginning at each shoulder and meeting just before the tail. The band resembles a saddle or an ancient harp. A fully matured male harp seal measures six feet in length and weighs nearly 400 pounds. Some live to be 30 years of age or more.

The hood seal (*Cystophora cristata*), about which we know much less than the harp, ranges from beyond Spitsbergen in the north to perhaps as far south as the Grand Banks. The three main breeding areas are Davis Strait, the West Ice north of Jan Mayen, and the Front, with a lesser breeding area in the Gulf of St. Lawrence. From late May until early June, the "Newfoundland" hoods feed off the west coast of Greenland, after which they migrate around Cape Farewell into Denmark Strait to moult in July and August. After moulting, a minority of the Newfoundland hoods return to the west coast of Greenland, penetrating northern Baffin Bay. Like the harps, the hoods move south at the onset of winter. Their route parallels that of the harp seals, the hoods keeping farther out to sea, that is, to the northeast. The hoods arrive at the Strait of Belle Isle in late December and divide there, part entering the Gulf of St. Lawrence and the remainder continuing down the northeast coast of Newfoundland. Once in the gulf, the hoods situate themselves south of the gulf harps. Hoods start their northward migration in time to reach the whelping ice in March, and give birth during the second half of the month. Lactation, which lasts only four days (the shortest of any mammal), is followed by mating. Then the adults resume their northward journey, most reaching the west coast of Greenland by late May.

The newborn hood pup is known as a "blueback." Its back is a slate-blue colour, its sides and belly a silver-grey shade. The blueback actually yields more oil and has a larger pelt than the whitecoat, but because bluebacks were so few and less accessible, the hunt focussed on whitecoats.[3] The pelt is also superior because the hood pup loses its fetal coat before birth and therefore does not go through a raggedy-jacket stage. Bluebacks take to the water around the first week of April. Until they reach maturity, young hoods are also known as bedlamers, although the term is used more in connection with harps. Spots begin to appear on the hood's coat at the age of two, and as an adult it will be grey with dark brown spots over all of its body. Hoods mature at three years. The adult male hood ("dog") weighs in the neighbourhood of 900 pounds and measures nine to ten feet from head to tail. The normal life span is around 20 years. The hood gets its name from the inflatable sack or hood attached to the dog's nose. Unlike most harps, adult hoods will try to defend their young.

The Ice

The ice is the seal's main defence. The extent of the ice field varies in any given year: a recent estimate places the area of the whelping ice at between 7 and 77 square miles. The ice is in constant motion, normally tending south or southeast at about a couple of knots per hour. The direction and extent of the ice depend on temperature and ocean currents, and especially on the wind. When the wind blows from the land, the ice is loose and easier to navigate; there is usually open water between the coast and the landward edge of the ice. This stretch of water, known as the "inside cut" or the "cow path," is the best route to the whelping ice. If the wind blows towards the land, the ice becomes compacted, closing the inside cut. Since harp seals prefer "local" ice (ice formed off Labrador) to the heavier arctic ice farther out to sea favoured by the hoods, access to the inside cut was crucial to the success of the hunt. On the other hand, the landsman hunt prospered when the wind blew on shore, often bringing the seals within easy walking distance. In the spring of 1862 the wind blew on shore for 52 days, driving the herd into Green Bay where, it was said, "even the women and dogs made £10 a man."[4] Thereafter, whenever a similar phenomenon occurred anywhere along the coast, it was called a "Green Bay spring."

Commodities

The rise of commercial sealing in Newfoundland coincided with the rise of industry in Great Britain. Before that, seals had been utilized by Newfoundland's native peoples and by the early European settlers. They burned seal oil for heat and light, ate the meat, and made the skins into boots, mittens, caps, and outer clothing. When commercial sealing began, only the oil and skins were of any value.[5] The oil was used in making soap and as a lubricant for industrial machinery. It provided light in mines, lighthouses, and homes until gradually replaced by coal-gas and kerosene as the 19th century progressed, although some lighthouses continued to burn seal oil until the early 20th century.[6] Seal oil was also sometimes used as an ingredient in high explosives. Tanned sealskins have been used in a variety of products, especially those requiring fine or patent leather, sealskin being thinner than many animal hides. The first bicycles were equipped with sealskin saddles and saddle bags. Skins were made into cigar cases, hand bags, wallets, boots, shoes, harnesses, portmanteaus, and bookbindings.

In the 20th century, seal oil found a new use after the development of the chemical process of hydrogenation (turning oil into fat by saturating its unsaturated acids with hydrogen). Thus seal oil, like whale oil, came to be used in the production of margarine and, later, chocolates. The controversial use of seal pelts in the manufacture of coats dates from the 1920s.

Of Men and Seals

Farley Mowat has characterized the Newfoundland seal hunt as "an organized exploitation of both men and seals."[7] He was, of course, referring to the commercial seal hunt, which dates from the early 18th century. For thousands of years before then, Newfoundland's native peoples hunted seals

16

not for profit, but for survival. And for many Newfoundlanders, from the early European settlers to their descendants of today, the same justification prevails, for seals, along with fish, animals, and birds, provide sustenance. Seal hunting was necessary not only because of the barrenness of the land, but also because the chief industry, the fishery, offered so little financial reward that if people ate only what they could afford to buy, they would have starved to death. It is in the area of commercial sealing, particularly the vessel-based hunt, that Mowat's description is valid. There, poor pay was combined with deplorable living conditions and an unsafe work environment. This picture did not improve substantially until the 20th century, and then most of the progress was realized only after the Second World War.

The sealing industry was of profound economic importance to Newfoundland in the 19th century and continued to have limited regional importance in the 20th. But the story of sealing is also a story of missed opportunities. The firms that came to dominate the industry in the 19th and early 20th centuries were based in Great Britain. It was there and elsewhere in Europe that most of the profits were realized. The British firms in turn lost control of the industry in the 20th century to Norwegian interests. This paralleled developments in the staple salt cod fishery, in which government and entrepreneurial neglect combined to topple Newfoundland from its leading international role.

The second great theme of the Newfoundland seal hunt is the exploitation of the seals themselves. Catches were already declining by the second half of the 19th century. The Newfoundland government enacted legislation aimed at easing the pressure on the resource, but the Great Depression and two world wars proved more effective. The harp and hood seal stocks were rejuvenated by 1945 and probably could have sustained prewar catch levels indefinitely. Unfortunately, postwar hunting was much more intense, partly because of increased hunting efficiency, but also on account of fierce Norwegian competition. Concern over the seals' survival, manifested by Canadian government scientists in the 1950s and by environmentalists a decade later, brought about much-needed regulation of the hunt in the form of catch quotas.

In 1983, years of opposition to the hunt culminated with the European Economic Community's boycott of whitecoat and blueback products. This was more than an isolated victory for anti-sealing groups; it also was a triumph of the modern way of viewing nature.

Origins and Early Development

Native Peoples

The earliest seal hunters in Newfoundland were its original native people, the Maritime Archaic Indians. Probably descendants of the Palaeo-Indians of the present-day Maritime provinces, they had crossed over to the north bank of the St. Lawrence estuary around 9000 years ago.[1] Artifacts from Maritime Archaic sites at L'Anse-Amour, Labrador, and Port au Choix, Newfoundland, show that seals were a dietary staple. Although no evidence remains, they likely used sealskin for clothing. Harps and hoods, along with less numerous harbour and grey seals, would have been available in the Gulf of St. Lawrence during late winter and spring; off Labrador, harps and hoods were complemented by ringed and bearded seals. The Maritime Archaic Indians may have timed their return from the interior, where they likely spent the winter living off caribou and lesser game, to coincide with the harps' arrival in the Gulf of St. Lawrence.

The L'Anse-Amour artifacts indicate that the seals were killed with barbed and toggled harpoons. The tip of the barbed harpoon detached from the shaft once it entered the seal's body and the barbs held it in place as damaged tissue expanded along the incision. The tip was attached to a hand-held line so the seal could be towed ashore or onto the ice. The tip of the toggled harpoon stayed in place by twisting or "toggling" in the wound.

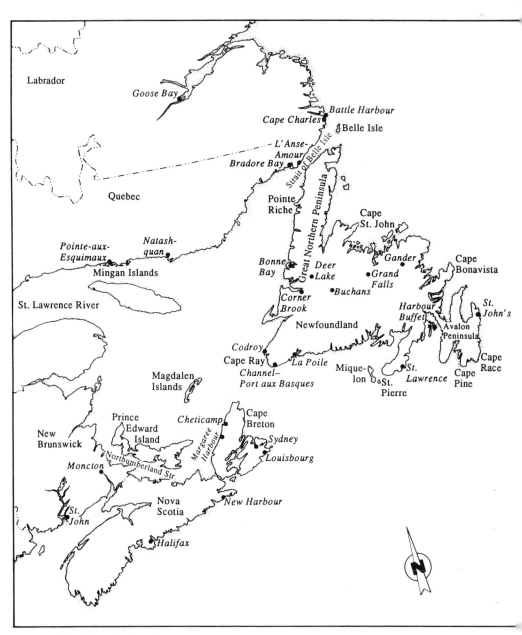

7 Atlantic Canada.
Map by Dorothy Kappler.

Seals figured prominently in the daily existence of the next native people, the Early Palaeo-Eskimos, who arrived in northern Labrador from the Canadian Arctic about four thousand years ago. They do not appear to have spread into Newfoundland. The Early Palaeo-Eskimos relied more heavily on coastal resources, since they had little contact with the interior. They adapted to year-round occupation of the coast by developing ice-hunting techniques that were more sophisticated than those the Maritime Archaic Indians used. The most notable technique was breathing-hole hunting, used primarily in hunting ringed seals. The hunter would locate the seal's breathing hole, wait beside it, and then harpoon the seal when it came up for air.

The third native group, the Dorset or Late Palaeo-Eskimos, may have descended from the Early Palaeo-Eskimos. Although the Dorset persisted until at least AD 800 on the island of Newfoundland, they became extinct in southern and central Labrador at an earlier date. They resemble the Early Palaeo-Eskimos in that they appear to have spent the whole year on the coast. The dispersal of the Dorset over a wide geographical area, from Labrador to southern Newfoundland, meant regional variations in consumption patterns. For example, in Newfoundland and southern Labrador, ringed seals were not available, so harp seals were hunted. Whales were taken throughout Dorset territory, as were numerous species of fish. Other staples were caribou (especially in Newfoundland), beaver, fox, walrus, and birds. The Dorset, too, used toggled harpoons in hunting seals.

The presence of soapstone lamps at Dorset Eskimo sites has led archaeologists to conclude that seal or whale oil was burned to produce heat and light. The importance of seals is further shown by charms and amulets found at most Newfoundland and Labrador native sites. The seal forms a recurrent motif on the charms and amulets, likely worn either to endow the wearer with luck in the seal hunt or to impart to him some of the favourable attributes of his prey.

The Dorset Eskimos were replaced in central Labrador by Indians who arrived fairly recently, perhaps only a thousand years before the first Europeans. These Indians may have been ancestors of the Naskapi and Montagnais Indians, who were present by the time of European contact. Although the Naskapi and Montagnais did at first spend part of the year on the coast, by the time of European contact they confined themselves wholly to the interior, living on fish and game during the short summer months and depending largely on the fall caribou hunt to get them through the winters.

The Thule or Labrador Eskimos, contemporaries of the Naskapi and Montagnais, did retain a migratory existence. The Labrador Eskimos, who call themselves Inuit (true men), arrived in northern Labrador from the Arctic *circa* 1300. Their rapid spread into central Labrador in the 16th century was probably a result of the trading opportunities provided by early European whalers and fishermen. It is unclear from the archaeological evidence whether the last of the Dorset culture were replaced or absorbed by the Labrador Inuit. It has been said that the entire economy of the Labrador Inuit depended on the seal,[2] and this is not far from the truth. Seal meat was a vital food source, and even the blood was boiled with other ingredients as a soup.[3] The oil was burned to provide light and heat, and was used to soften dry food. Sealskin was the raw material for footwear, mittens, caps, and outer clothing. It was also used to cover tents or huts framed out of whalebones. The seal was therefore a source of food, shelter, and clothing. As to hunting methods, the Labrador Inuit attached inflated bladders to a harpoon line to create drag as a wounded seal attempted to escape.

Seals became even more important to the Labrador Inuit after the early years of white contact when Europeans and New Englanders decimated the whale population. Under the pressure of the white man's advance, the Labrador Inuit withdrew farther and farther north. They were also harassed from the interior by the Naskapi. Threatened on two flanks, they retreated to the Moravian missions of northern Labrador in the 19th century, where they continued to hunt and fish under the aegis of the Moravian brethren.

On insular Newfoundland the importance of seals to coastal habitation was also reflected by the Indian cultures that appear once again in the archaeological record as early as AD 200 at Cape Freels. Co-existing with the Dorset for several hundred years, these Indian cultures developed several distinct traditions known archaeologically by such names as Little Passage Complex and Beaches Complex. Of these, the Little Passage Complex seems to be the most likely direct ancestor of the historically known Beothuck culture. The Cape Freels site seems to have been chosen for hunting grey and harbour seals, which can be taken in the area even today. Seal bones were discovered at a known Beothuck site, Wigwam Brook on the Exploits River, dated to the late 18th or early 19th century. But the overwhelming majority of the bone remains at Wigwam Brook were from caribou. This stems from the fact that the site, an inland one, was occupied year-round. White settlement had become so well established along the northeast coast that the traditional Beothuck migrations between the coast and the interior were totally disrupted. The loss of access to seals and other coastal resources was a key factor in the extinction of the Beothucks by the late 1820s, since the resources of the interior could not sustain them for the entire year.

The Europeans

The development of a full-fledged sealing industry by European immigrants to Newfoundland was presaged by "fisheries" based on other sea mammals in the area. Basque whaling began in the Strait of Belle Isle in the 1540s and peaked in the 1580s.[4] In the 1590s, Basques and Bretons killed large numbers of walruses on the Magdalen Islands. Together with Micmac Indians, they prevented a small group of Englishmen from settling on the Magdalens in 1597, no doubt to preserve their monopoly.[5] Although the Basques and Bretons probably did some incidental sealing, the settlers of New France were the first to take it up on a significant scale. In 1661 the Company of One Hundred Associates granted François Bissot a seigneury and fishing rights extending roughly from the Mingan Islands to Bradore Bay.[6] Bissot fished, sealed, and traded along the length of this coast. By the early years of the 18th century, sealing had become an established activity among *Canadiens* who visited the Labrador coast from October to June and then returned to their homes on the St. Lawrence's north shore.

It is generally believed that English-speaking settlers in Newfoundland learned their sealing techniques from the French, particularly from Jerseymen active in Labrador.[7] The industry sprang up in northeastern Newfoundland in the first half of the 18th century and before long was attracting settlers to the region, which had previously been neglected in favour of the Avalon Peninsula. Until the last decade of the century, sealing was not important south of Trinity.[8] Bonavista Bay was briefly the focus in the 1730s, but by the 1750s emphasis had shifted north to the Fogo-Twillingate area of Notre Dame Bay. After 1763, when the Treaty of Paris placed Labrador under the jurisdiction of the governor of Newfoundland, there was a surge of British interest in the region. Jeremiah Coughlan, a merchant with holdings at Fogo, founded the first British sealing post on the Labrador coast in 1765.[9] By the 1770s, seal oil produced at Labrador had an average annual value of £7000, compared with £1700 and £1200 for Fogo-Twillingate and Bonavista Bay respectively.[10]

Sealing was then a land-based, winter activity, the seals being caught in nets as they migrated southward. British naturalist Joseph Banks, who visited Newfoundland in 1766, wrote that the seals were taken exclusively during December.[11] A series of nets were placed in channels the seals frequented, either between the mainland and an island or between two islands or large rocks. The last net in each series was stretched tight while the others lay slack near the bottom. As soon as enough seals became entangled in the last net, the next net would be pulled tight in front of it, and so on until all

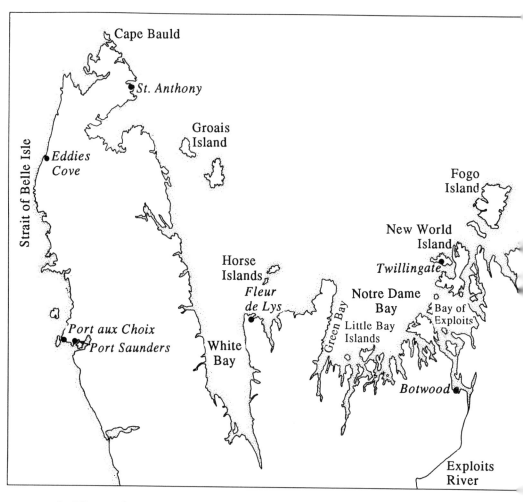

8 The northeast coast of Newfoundland.
Map by Wayne Hughes and Dorothy Kappler.

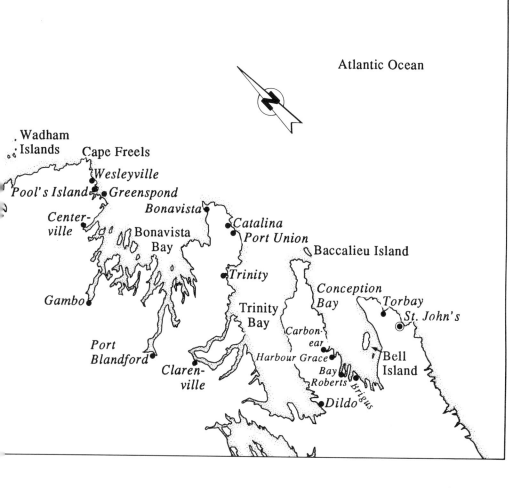

Atlantic Ocean

Wadham
Islands
Cape Freels
Wesleyville
Pool's Island
Greenspond
Center-
ville
Bonavista
Bonavista
Bay
Catalina
Port Union
Baccalieu Island
Trinity
Gambo
Conception
Bay
Torbay
St. John's
Trinity
Bay
Carbon-
ear
Port
Blandford
Harbour Grace
Bell
Island
Claren-
ville
Bay
Roberts
Brigus
Dildo

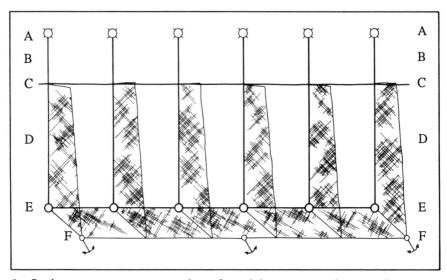

9 Seal nets. AA, capstans on shore for raising nets; BB, hawsers from capstans to nets; CC, water's edge; DD, nets extending from water's edge seaward to the outer net; EE, outer net; F, anchors.

After Edward Chappell, *Voyage of His Majesty's Ship Rosamond to Newfoundland and the Southern Coast of Labrador....* (London: Printed for J. Mawman, 1818), p. 198.

the nets were raised, creating a series of pounds. The seals drowned as they tried to escape through the netting. This was the preferred method, but few places afforded perfect natural conditions, so the men often resorted to another method, which magistrate John Bland of Bonavista described in 1802.[12] The nets were anchored in open water, sometimes in depths of 40 fathoms, by means of killicks (stones encased with wood).[13] The top of each net was buoyed up by corks, and the ends held by poles that stood upright in the water. Thus the nets were said to "stand on their legs." Four or five men normally worked 20 nets, the crews being doubled if there was a heavy run of seals. Contrary to Banks, Bland observed that sealing began around Christmas and continued through the winter.

Residents of Labrador took netting technology one step further with the development of a seal-net frame.[14] The series of nets Banks had described was now complemented by a long net that ran parallel to the shore at the seaward ends of the individual pound nets. Now there was little way the seals could escape the pounds. In Labrador sealing began in November and ended around Christmas, when the bulk of the seals had passed to the south. After the seals were brought ashore, they remained frozen until late April or early

May when they were thawed in the sun.[15] Then they were skinned and the fat separated from the pelts, chopped up, and fed into iron boilers to be rendered into oil.

Towards the end of the 18th century there arose a more daring method of killing seals. Crews began to venture out into the bays aboard shallops, from which they shot the seals.[16] The shallop was a wooden vessel 30 to 40 feet long and decked fore and aft, with movable deck boards amidships where the seals were stowed.[17] It could be either rowed or sailed and possessed mainmast, foremast, and lugsail. The usual crew was three or four men who took shelter in "cuddies" at either end of the vessel. Shallop crews did not go onto the open sea but stayed inside the headlands of the major bays. The significance of the shallop, aside from the greater range it provided, was that it allowed the seal hunt to be conducted during the spring, just after the seals' breeding season. Before, sealing had been confined largely to the winter except whenever spring ice carried the seals within walking distance of land. Shallops were an important link between the landsman hunt and the vessel-based hunt that was to emerge in the 1790s.

The Seal Hunt
1793–1861

Overview

The seal hunt began a new phase in 1793 when a St. John's merchant sent two schooners of approximately 45 tons each "to the ice."[1] Others soon followed suit. In 1799, 120 000 seal pelts (the skin with the fat still attached) were landed in St. John's and Conception Bay, two areas that had not even participated before 1793.[2] The greater range afforded by the schooner meant that living near the seal herds was no longer the all-important factor it had once been. Thus, during the course of the 19th century, St. John's and Conception Bay, with their greater populations and financial resources, came to dominate the hunt. Lesser numbers of schooners sailed from northern bays, where the landsman hunt now co-existed with the vessel-based hunt. The adoption of schooners completely transformed sealing in Newfoundland, both in its methods and in its scale of operation. Sealing came to rival the cod fishery as a pillar of the Newfoundland economy. But the industry's rapid expansion was based on an incredible assault on the seal herds that ultimately damaged the industry itself.

During the course of the Napoleonic Wars (1793–1802, 1803–15), exports of seal products ranged in value from £20 000 to £60 000 annually, compared with annual exports of salted cod worth approximately £600 000.[3] The expansion of sealing was aided by the simultaneous development of the

29

northern fishery.[4] This fishery was born when Newfoundland fishermen pushed up the northeast coast into the French Shore, extending from Cape St. John to Cape Ray, which French fishermen had vacated because of the wars.[5] Many of the vessels used were diverted from the Grand Banks fishery, which had also been disrupted by the wars. A system evolved whereby the same men and vessels that sailed to the seal hunt in the spring also took part in the northern fishery in the summer and fall. This maximized the use of the vessels; they were now active during all months when navigation was possible along the northeast coast. When French fishermen returned to the French Shore after 1815, the northern fishery shifted to Labrador. Throughout the remainder of the 19th century, sealing and the Labrador fishery maintained their close links.

After the Napoleonic Wars the sealing industry continued its amazing growth. By 1844, seal products accounted for over one-third of the value of all Newfoundland exports.[6] Even though that tapered off to 25 per cent by mid-century, capital investment in the industry reached a historic peak in 1857 when 370 vessels and 13 600 men sailed to the ice, taking over 500 000 seals worth £425 000.[7] It has been estimated that in the period 1800–1860, the Newfoundland sealing fleet landed "at least" 18 million seals.[8] The most prolific decades were the 1830s and 1840s, with the most seals ever taken in one year being 687 000 in 1831. By the mid-fifties, however, decline had begun.

The social and economic impact of sealing was profound. In the 18th century the landsman hunt, allied with the salmon fishery and fur trapping, had provided a basis for year-round settlement in areas of the northeast coast previously dependent upon the summer cod fishery.[9] Settlement received a further boost after the development of the vessel-based hunt for seals in the 1790s. Before 1840, the wooden vessels used in sealing were all built in Newfoundland.[10] Although after that date the larger vessels tended increasingly to come from elsewhere in British North America, schooners were still built locally. This native shipbuilding industry provided winter employment for carpenters, sailmakers, blacksmiths, and riggers, both in building new ships and in repairing old ones. In addition, each major sealing centre had its own facilities for preparing skins and manufacturing oil. Sealing therefore fostered population growth and freed the economy from its nearly total dependence on a single industry, the cod fishery.

Most of the wealth the sealing industry generated was accumulated on the Avalon Peninsula and particularly in Conception Bay. In 1833 Conception Bay ports sent 205 ships and 4526 men to the ice.[11] St. John's was next with 110 ships and 2536 men, while all other Newfoundland ports accounted for

44 ships and 921 men. In 1861 Conception Bay still led St. John's, accounting for 152 vessels and 7420 men, compared with 73 vessels and 3836 men from the capital.[12] Communities north of Conception Bay only slowly shifted to the vessel-based hunt because they were still active in the landsman hunt. Trinity, although blessed with an excellent harbour, sent a mere 12 sailing vessels to the ice in 1845. This grew to 16 in 1857 and 29 in 1869.[13] The landsman hunt was especially popular in the Fogo-Twillingate area, and to a lesser extent in Bonavista Bay,[14] where the hunters' advantage was their proximity to the seals' migration routes and breeding grounds. The landsman hunt required only a modest capital outlay for nets and boats, yet it could yield high profits in good years. Nevertheless, the vessel-based hunt had the greater overall impact on the Newfoundland economy, providing employment for more men, directly and indirectly, and requiring substantially greater capital investment.

Preparations

Preparations for the seal hunt began with awarding berths to men who wished to go to the ice. This traditionally took place on Boxing Day at the business premises of the merchants who owned the vessels.[15] Shortly after Candlemas (2 February), the actual work started.[16] The chief task was cutting and hauling timber for spars and for building new punts, the small boats the sealing ships carried. During this period the sealers received food but no pay.

The vessels had to be modified for ice navigation. The hull was protected by a covering of greenheart or local hardwood such as birch, and this in turn was encased in iron sheathing.[17] The hull was also strengthened by adding extra beams, usually a single row eight feet apart at the water line for smaller vessels, and two rows on larger vessels, one at the water line and another immediately below the deck. By 1830, wooden pounds were commonly built in the hold to prevent the seal pelts from shifting. Ballast, usually rocks, was placed in the pounds, to be removed during the course of the voyage to make way for pelts.

Vessels

The trend from the 1790s through the 1850s was towards increasingly larger vessels. In 1819 most sealing vessels were schooners in the 40- to 75-ton range, as well as decked boats, probably shallops, of 24 to 35 tons.[18] But 1819 was also the year when the first vessel of over 100 tons went to the ice, the 104-ton brig *Four Brothers*, owned by William Munden of Brigus.[19] Twenty-one years later the average sealing vessel was a schooner, brig, or brigantine ranging from 80 to 150 tons; by 1857 some vessels were over 200 tons.[20] Most schooners had the usual fore and aft rigging, and most brigs the usual square sails on both masts, but variations were designed to allow greater manoeuvrability in the ice. The "beaver hat man," similar to a topsail schooner, was a 60- to 80-ton schooner with fore and aft rigging complemented by a square topsail and yards on the foremast.[21] The "jackass brig" of 100 to 140 tons had a square rig on the foremast, a square topsail, and a topgallant sail on the mainmast.

Season Length

In the early years of the 19th century, sealing vessels left for the ice around St. Patrick's Day, although some waited as late as 21 March to avoid the equinoctial gales.[22] By 1828, however, it was common for sealing vessels from Carbonear and, one assumes, the rest of Conception Bay, to have sailed by St. Patrick's Day.[23] Before long, most vessels sailed by 1 March or shortly after.[24] There was a good reason for the trend towards earlier departure dates: vessels that did not sail until the middle of March largely missed the valuable harp seal pups, who began taking to the water around the end of the month.

The first trip to the ice lasted four to six weeks.[25] If a vessel finished early with a full load, it might make a second or even a third trip. As a rule, all vessels were back in port before the end of May so they could be made ready to sail for the Labrador fishery during the first week of June.

Navigation

The earlier departure date meant that ice navigation became increasingly difficult. Sometimes vessels were frozen into their own harbours. When this happened the men used ice-saws to cut a passage through the ice, singing shanties while they worked.[26] Once a passage was cut, the crews used hawsers to pull their vessels through the ice. It was not an easy task. In Brigus, where the harbour was over a mile long, it sometimes took more than ten days to get the fleet on its way![27]

Most vessels sailed north through the "inside cut," but sooner or later they had to enter the ice field to locate the seals. Ideally the vessels followed leads in the ice, but leads often closed up or simply did not exist. Many sealing vessels were equipped with rams, two long poles attached to opposite sides of the bow and lashed together just below the bowsprit.[28] The men hung onto the rams and tried to break the ice with their feet. Or they hung by ropes from the bow to stamp at the ice.[29] Another popular device was the poker or stabber pole, a wooden pole 12 to 15 feet long used to pound and break the ice. As pieces of ice were broken off, they were levered under the adjoining ice. In thick ice the crews resorted to ice-saws, axes, and chisels, again pulling the vessel through the passage by means of a hawser. Sometimes, as well, grappling hooks were fixed to the ice in front of the bow and the men tried to warp the ship along. If a vessel became jammed, sticks placed along its sides absorbed some of the pressure. In spite of this precaution, many vessels were crushed by the ice.

Crews

In 1819 the largest schooners had crews of 13 to 18 men.[30] As the vessels got bigger, so did the crews, and eventually the larger vessels carried 40 to 50 men.[31] In 1840 the 36 crewmen of the 120-ton brigantine *Topaz* were formed into three watches of 12, and nine four-man punt crews.[32] Each watch was headed by a master watch appointed by the captain. The master watch acted as mate while his watch was on shipboard duty, and became officer in charge whenever the captain was below. The three master watches commanded three of the punt crews, and the captain appointed another six men to command the others. These nine selected their own crews. On some vessels an entire watch was needed just to work the sails.[33] A scunner or barrel-man was chosen from each watch to guide the vessel through the ice, which he did from a crow's-nest atop one of the masts. The men in each watch tended to be from the same community, and members of the same family commonly stayed together.[34]

Equipment

The men killed seal pups with gaffs or bats. The gaff was a wooden pole made of dogwood or spruce, five to eight feet long with an iron hook at one end.[35] It was also an important safety device: it was used for testing the ice, for steadying men jumping from pan to pan, for hauling men out of the water, and for helping men scramble aboard the vessel from the ice. In the sailing era the men appear to have supplied their own gaffs.[36] The bat was a tapered club approximately three feet long. The gaff was the preferred instrument, probably because of its greater all-round utility. Some authorities claim the gaff's popularity derived from its being a "more efficient killing implement" than the bat.[37] However, 19th-century accounts indicate that many sealers did not care whether the seal was dead or not.[38]

Each sealer possessed a sculping knife, a broad clasp knife used to remove the pelt from the carcass.[39] Around his shoulder he carried a tow rope, used to haul the pelts back to the vessel. Although details of the tow rope are scanty for this period, later versions were 12 to 14 feet long and made of half-inch manila hemp.[40] The tow rope of the sailing era was slightly tapered at one end so it could be passed easily through slits cut in the pelt.[41] For traction on the ice the sealers wore boots ("skinny woppers"), often made of sealskin, with studded heels and soles. The studs were called "sparables," "chisels," or "frosters."[42]

Punt crews required different equipment, the most obvious being the punt itself. Unfortunately, we lack a good contemporary description of the sealing punt. We do know that it had four oars, a crew of four men, and a capacity of 30 unskinned ("round") adult seals or 60 pelts.[43] Additional seals were sometimes towed behind the punt. Because they sought mainly beaters and adult seals, punt crews did not kill with gaffs or bats but used guns called "long-toms."[44]

> *The sealing-gun is an immense affair, as long as a duck-gun, but with a much wider bore, roughly made, and, in some instances, not over-sound. The men put in a great charge of powder and shot — frequently ten fingers' breadth, or even more — the powder being coarse, and the shot larger than buck-shot, consisting, in fact, rather of small bullets than shot, being cast, and not dropped. It was as much as I could do to hold one of these guns straight out; and the men were frequently knocked down by its rebound, when they fired standing on slippery ice.[45]*

The Whitecoat Hunt

The first stage of the hunt was the whitecoat hunt. The sealers' deployment on the ice depended upon the shape of the seal patch, the proximity of other vessels and, most important of all, ice conditions.[46] The ideal arrangement was for the men to radiate outward in a circle from the vessel, but this was seldom achieved. Watches were subdivided into groups of two or three men, both for safety purposes and so they could help each other in various tasks, such as hauling a particularly heavy "tow" of pelts.

Ordinarily, in the whitecoat stage the sealers travelled by foot on the ice, a relatively straightforward activity if the ice was unbroken. When the ice was broken, the sealer required great skill to jump from pan to pan, a practice known as "copying." Sometimes the ice was so loose that substantial stretches of water lay open and punts had to be used to cross the leads and transport pelts back to the vessel.

The sealer killed a whitecoat by striking it twice on the nose with his gaff, stunning it with the first blow and killing it with the second.[47] He then removed the pelt or "sculp" with his sculping knife, cut two slits in it, and passed the tapered end of his tow rope through the slits and the eyeholes. As pelts were accumulated, they were laid one upon the other, joined by the tow rope. A "tow" consisted of anywhere from three to seven pelts depending on the size of the whitecoats, five or six pelts constituting a heavy load. The average pelt was 3 feet long, 2 1/2 feet wide, and weighed 35 to 40 pounds.[48] After assembling a tow of pelts, the sealer fastened his gaff to the tow, threw the tow rope over his shoulder, and hauled the pelts back to his vessel. Occasionally, such as when punts were used, dead whitecoats were brought back "round" to the vessel and skinned on the deck at night.[49] The carcasses were thrown overboard and the pelts left on deck for a couple of days to freeze, which reduced the tendency of the fat to run to oil once the pelts were stowed in the pounds below deck.[50] As the pelts were put below, ballast was removed from the pounds and thrown overboard.

The Beater and Adult Seal Hunts

If enough whitecoats were killed to fill the hold, the vessel would return to port at once, unload, and set sail again to hunt beaters and adult seals. Given the less than perfect manoeuvrability of the sailing vessels, most did not get a full load of whitecoats and immediately went on to hunt beaters and adults.

The worst possible situation was one in which hardly any whitecoats were taken. Then the beaters and adults were essential to save the voyage.[51]

During the beater stage of the hunt, guns replaced gaffs. Also, punts were more likely to be required because beaters were hunted towards the end of March when the ice was more liable to break up. Each gunner was accompanied by two helpers, known as "dogs," who carried his shot and gunpowder. The gunner hid behind a pressure ridge in the ice and attempted to shoot the seal in the head, which caused instant death and did not damage the hide.[52] The dogs were responsible for skinning any seals the gunner killed.

The hunt for adult hoods and harps constituted the final stage of the season and incorporated the same methods used in the beater hunt. The adult hunt did not get under way until well into April. As with the beaters, adults were hunted to save a voyage that had taken few whitecoats, or else they were the focus of a second or third trip by a successful vessel. Sealers said that the key to killing hoods was to shoot the female first. The male would not go far from his mate and therefore made an easy target.[53]

Both the beater and adult seal hunts were extremely wasteful. Seals shot in open water frequently sank before punt crews could reach them. According to an experienced sealing captain, only one adult seal was taken for every 20 that were lost.[54] Also, because of its breeding capacity, each adult female "was potentially equivalent to fifteen whitecoats slaughtered in the early phase of future hunts."[55] Given the scale of the beater and adult seal hunts in the sailing era, these hunts were undoubtedly influential in reducing seal stocks, a reduction that became apparent by the second half of the 19th century.

Health and Safety

Danger and squalor were bywords for the Newfoundland seal hunt. The wooden sailing vessels were at nature's mercy, and the toll it took from them was staggering. Although there are no accurate records, it has been estimated that over 400 sailing vessels were lost during the 19th century.[56] In 1832 a hurricane claimed 14 schooners and the lives of over 300 men.[57] In 1852, 40 vessels were crushed in the ice near the Wadham Islands, but fortunately all 1500 sealers were able to escape to safety on the islands, where they were later rescued by a government relief vessel.[58] In 1864, 26 vessels were crushed and another 140 were jammed in the ice for several weeks at the mouth of Green Bay. Approximately 1500 sealers made it to Greenspond, where they received emergency provisions brought by a government-chartered brig. In 1871 the schooner *Huntress* of Bay Roberts and its crew of 40 men were lost at the ice. The next year 43 more men were lost when the brig *Huntsman* was wrecked off Cape Charles, Labrador.

Not only ice pressure destroyed vessels. Sometimes small icebergs known as "growlers" littered the water. Growlers barely protruded above the water and were notoriously unstable, turning over suddenly if their equilibrium was disturbed.[59] This phenomenon plagued the fleet in the 1844 season, thereafter referred to as "the spring of the growlers." Sealers frequently became separated from their vessels during sudden snowstorms. In 1873, 24 men out in the schooner *Deerhound*'s punts became separated from their vessel in a storm and perished. The chances of storms are particularly great in the North Atlantic in March, for this is the time of the equinoctial gales (the sun crosses the equator on 20 March). Sealers called these storms "brushes," the two main ones being Paddy's Brush (17 March) and Sheila's Brush (18 March), after St. Patrick and his legendary helpmate. Fog was another hazard. Various signals guided men back during fog. The brigantine *Topaz* fired a nine-pounder carronade twice every 30 minutes.[60] Other signals were not as sophisticated. They included banging a frying pan with a poker, clapping pieces of wood together, and thumping a rope against the bulwarks.

Living conditions on the vessels were despicable. The men were crammed into the forecastle, sarcastically named "the ballroom."[61] A captain of a larger vessel did have an after-cabin. Food consisted of ship's biscuit (hard bread), tea, salted fish, and salted pork. It is hardly surprising that once the vessels were into the seals, they became part of the men's diet, particularly the livers, hearts, kidneys, and flippers. In 1840 "the constant employment of the men on deck, when they had nothing else to do, was boiling, frying, or roasting pieces of seal flesh and eating them. Immediately after a dinner or breakfast down below, they would come on deck and set to work at the seal by way of dessert."[62] The men sometimes ate the hearts and kidneys raw, since they believed this prevented scurvy.[63] Towards the end of the 1840 season most of the food on the *Topaz* had been consumed; little more than tea and rum remained.

Cleanliness was out of the question after the first seals had been brought aboard. Vessel and crew became covered in blood and grease, giving rise to the name "greasy-jacket" to describe a sealer. As the *Topaz* returned to port, it "was perceived to stink most awfully, and everything on board, including the bulk-heads of the cabin, began to sweat with grease."[64]

Processing

When the seal pelts were unloaded, the skinners were the first to take over. Their job was to separate the skin from the fat. The skins were then stretched, salted, and stored for export, bundles of five being common.[65] The biggest job was rendering the fat into oil, a process that lasted up to two months.[66] The fat was cut into small pieces and fed into huge vats of 15 to 20 tons' capacity where it was melted by the sun's warmth. The vats were square, built of timber, tarred inside and outside, and reinforced with iron at the corners and bottoms. Through openings at various heights, controlled by plugs or faucets, the oil was drawn off and collected in hogsheads. The best oil was at the top, the quality decreasing at each lower level. When all the oil was drained from the vats, the remaining sediment was boiled in copper cauldrons to produce yet more oil, albeit of inferior quality. In 1861, 100 per cent of Newfoundland's 375 282 sealskins and 94 per cent (4957 tons) of its seal oil were exported to the United Kingdom.

Pay

To understand financial arrangements in the Newfoundland seal hunt of the 1790s through the 1850s, one must know something of the island's social structure. There were three main classes.[67] The merchant class, based in Great Britain, possessed branch establishments in St. John's and in the larger outports. The "planter" class was a middle class between the merchants and the fishermen, who formed the third class. Planters bought or rented vessels from the merchants, taking the vessels to the seal hunt in the spring and to the Labrador fishery in the summer and fall.[68] The best account of the inter-action of the three classes comes from Philip Henry Gosse, a clerk at the Carbonear premises of the British firm Slade, Elson, and Company in 1828.[69] In Carbonear there were about 70 planters among a population of approximately 2500. More than one-third of the planters dealt with Slade, Elson, and Company. Prior to the start of the seal hunt in March, the firm provided each planter, on credit, with the necessary provisions and equip-ment (gunpowder, shot, punts, etc.) to enable him to outfit his schooner. The planter then selected ("shipped") a crew, whose names were registered with the firm. After registration the crew were free to draw goods on credit from the firm to the value of one-third to one-half of their probable earnings. This advance, known as the sealers' "crop," covered not only sealing gear, but also food, clothing, and other necessities the men and their families required. For all items advanced on credit, Slade, Elson, and Company charged about double the price obtaining in England. Goods purchased with cash were not marked up quite so steeply, but few fishermen had cash.

The accounts were squared at the end of the voyage, everything depending on the price the merchant set for the pelts. Up to 1835, most pelts were purchased by count; that is, by a certain amount per pelt.[70] However, this system was abandoned in 1836 because sealers abused it by cutting off parts of the pelts in order to lighten their tows. Weight then became the sole determinant. The proceeds of the voyage were divided into shares, half to the crew and half to the planter if he owned the vessel. If he did not, the merchant paid him a prearranged amount by count or weight, ranging from fourpence to sixpence per seal if by count.[71]

From each man's share the merchant deducted the value of his crop, plus the amount of the berth fee, a sum paid for the right to serve on a sealing vessel. In 1819 the berth fee was approximately 40 shillings.[72] By 1848 it ranged from 10 to 35 shillings, although at that date each man also had to provide 25 pieces of firewood for fuel.[73] The amount of the fee depended on the size and quality of the vessel. In 1840 the berth fee for the brigantine *Topaz*, a "superior" vessel, stood at £4 (Newfoundland currency was approximately five-sixths as valuable as British sterling).[74] One can safely assume that vessels with berth fees in the ten-shilling range were the smallest schooners. Since gunners usually provided their own guns, they did not have to pay a berth fee.

Sealers were paid partly in cash and partly in goods — the firm issued vouchers to sealers that they could exchange for goods at the firm's store. Sealing was the only branch of the Newfoundland fishery in which the worker received any cash for his labour. In 1832 some five thousand sealers from Harbour Grace and Carbonear struck for the payment of all sealing accounts in cash.[75] The strike was a success, and from that date Newfoundland vessel-based sealers were paid in cash only.

How much did the sealers make? In 1805, shares ranged from £5 to £25.[76] Fourteen years later each share brought an average of £9 to £12 sterling per man.[77] Thus two "bumper" trips in 1819 might put a sealer in the upper range for 1805. From his share had to be deducted the amount of his crop as well as his berth fee. This assumes that all sealers made money. In truth, the vast majority ended up owing money to the merchants. In the early 1860s no more than 60 vessels a year made successful trips.[78] Out of a fleet of 292 vessels in 1860, only 20.5 per cent would have turned a profit. Nevertheless, the chance to earn cash was an irresistible lure in an economy otherwise based on credit.

Merchants and planters were also imbued with the gambling spirit. Although the enterprise was risky, returns were sometimes so high that in one good season several years of losses could be completely recovered. From 1837 to 1853 the value of seal products per ton of shipping employed hovered between £7.1 and £9.7.[79] By 1860 the figure had fallen to £5.4, and anxious owners began to look for new ways to increase catches and restore profits.

The Seal Hunt
1862–1939
An Overview

The Impact of Steam

In 1862 the Newfoundland seal hunt entered the steam age. That year, two Scottish steamers participated in the Newfoundland hunt as a complementary activity to the Davis Strait right-whale "fishery."[1] Davis Strait whaling had been born in the 1820s following the collapse of British whaling off Greenland, but it in turn succumbed to overharvesting.[2] By the 1830s the British northern whaling trade, centred in Scotland, was in decline, the number of vessels dropping from 91 in 1830 to 25 in 1843.[3] The city of Peterhead staved off economic depression by branching out into the Newfoundland seal hunt. Other Scottish cities picked up on the Peterhead example and began sealing as well, primarily off Greenland but also off Newfoundland. This was not a radical departure for the whalers, since seals had always been regarded as "a possible make-weight on unsuccessful whaling expeditions."[4] But the emphasis had clearly shifted from whales to seals: between 1848 and 1857, Scottish vessels took nearly a million seals.

Scottish whaling and sealing took on a new dimension in 1857 when a few Hull whalers were outfitted with steam engines. Observers were impressed with the increased speed and manoeuvrability of the vessels, and before long Scottish steamers were making two trips a season, the first to the Greenland sealing grounds and the second to the whaling grounds. A series of poor returns from Greenland sealing prompted the Dundee Seal and Whale Fishing Company to send two of its steamers, *Camperdown* and *Polynia*, to the Newfoundland sealing grounds in 1862. The owners proposed to unload the pelts at St. John's (so the vessels could avoid a return trip to Scotland) and make straight for the whaling grounds, thereby reducing time spent in transit and potentially increasing profits.[5] The experiment was a disaster because 1862 was the year of the Green Bay spring, when landsmen took most of the seals while the fleet was jammed in the ice. The *Camperdown* was one of the vessels that got stuck and did not get a single seal. *Polynia* took only four. Scottish shipowners quickly cooled towards Newfoundland and did not repeat the experiment until 1867 when the SS *Esquimaux* was sent out, but it also returned clean. The Scots next sent steamers to Newfoundland in 1876, again owing to a decline in the Greenland seal herds. This time they were successful, and by 1881 six Scottish steamers, with Scottish and Newfoundland crews, were sealing off Newfoundland. A short-lived whaling revival in the 1880s diverted attention from Newfoundland and the 1881 participation level was never exceeded. In the 1890s Scottish interest in the Newfoundland seal hunt waned further as stocks declined to the point where voyages to Newfoundland ceased to be profitable. No Scottish steamers participated in the Newfoundland seal hunt after 1900.

The examples of *Camperdown* and *Polynia* in 1862, although undistinguished, prompted an interest in steam among Newfoundland sealing merchants who had seen their profits decline in the 1850s. They were particularly struck by the manoeuvrability of the *Polynia*, which rescued the crews of two jammed sailing vessels. In 1863 Walter Grieve and Company and Baine Johnston and Company, Scottish firms with St. John's branches, each purchased a steamer for the seal hunt.[6] The success of the steamers in 1865 and 1866 inspired imitators, and in 1867 eight Newfoundland-based "wooden walls" (wooden steam vessels) took part. As revenues began to rise again (£11.7 per ton employed in 1878), the number of steamers rose to 18 in 1873 and to an all-time high of 27 in 1880 and 1881.[7] The 27 steamers in 1881 carried 5815 men and took 281 949 seals.[8]

The change to steam brought with it a shift in the balance of power within the industry from the outports to St. John's, a shift that had already begun with the move to larger sailing vessels. The costs associated with the steamers and new processing equipment were so high that only the largest firms, all based in St. John's, were able to survive lean years.[9] And in the second half of the 19th century there were plenty of lean years because the seal herds continued to be overharvested. For a while the outports attempted to keep up, and merchants in ports like Brigus and Harbour Grace had their own steamers. However, this was a short-lived phenomenon. By 1896 the entire steam fleet was based in St. John's.[10] The gap between St. John's and the outports further widened following the introduction of steel ships in 1906. Sailing vessels generally ceased to be important to the success of the industry. Sealers from areas north of Conception Bay operated longer under sail, but by 1884 even the once-thriving sealing port of Trinity sent only three sailing vessels to the ice.[11]

The decline of the sailing vessels was accentuated by the steamers' negative effects on the Labrador fishery. With the concentration of the sealing industry in St. John's, outport vessels remained idle during what had often been their most profitable period of use — the spring seal hunt. This undermined the economic viability of outport sailing vessels, forcing some of their owners to withdraw from the Labrador fishery as well. Outport vessel owners had also profited from annually transporting fishermen and their families to and from coastal Labrador, but there too the steamers interfered. Steamer owners saw passenger traffic as a means of reducing losses during the off-season, when many steamships remained idle in St. John's harbour. It is no coincidence that the Labrador fishery began to decline in the mid-1880s when the steamship was supreme in the seal hunt.[12]

The impact of steam was more profound and more immediate on the sealing and shipbuilding industries than it was on the Labrador fishery. Steamers were more productive than sailing vessels, requiring half as many men to take an equivalent number of seals. The 27 steamers in the fleet in 1881 carried 5815 men, compared with 292 sailing vessels carrying 14 121 men in 1860.[13] In the 1870s, in an attempt to revive shipbuilding and to give work to displaced sealers, the Newfoundland government sought to develop a fishery on the Grand Banks by means of a bounty system for locally constructed vessels. After some initial success in the 1880s, the bank fishery settled back to an average level of 99 vessels per year in the 1890s, or only one-third the size of the sealing fleet in 1860.[14] The bank fishery employed an average of only 1324 men a year during the 1890s, hardly compensation for the more than eight thousand sealers' jobs that had disappeared between 1860 and 1881. Clearly, it was no panacea for the ailing shipbuilding and sealing industries.

The Newfoundland economy of the period was further hampered by an overcrowded inshore fishery characterized by declining yields. Other than government relief, the only option was emigration. Between 1884 and 1901, net emigration from Newfoundland ranged from 1500 to 2500 people per annum.[15] The brunt of the damage was borne by the outports. The population of Harbour Grace, for example, declined from 14 727 in 1884 to 12 671 in 1901. It would be impossible to determine how much of this was due to reductions in shipbuilding, sealing, and the Labrador fishery, and how much was due to conditions in the inshore fishery. But there is no doubt that the advent of steam made important changes in the sealing industry.

What the outports lost, St. John's gained. By the end of the 19th century, eight St. John's firms controlled the entire sealing industry.[16] In the process of this economic transformation, the middle and upper classes in the outports were drastically reduced. Steam "was as bad for the merchants as for the men."[17] The outport merchants who survived became little more than agents for St. John's firms, which themselves remained branches of British corporate empires, A.J. Harvey and Company being the chief exception. Many of the great outport sealing captains, who ranked just below the merchants in the economic hierarchy, drifted into poverty with the passing of the sailing fleet.[18] Sealing had passed from the age of competition to the age of monopoly.

The second major trend of the prewar period was the decline of the seal resource. Seal herds had been overharvested in the sailing era, and overharvesting became more intense after mid-century. From the highs of the 1830s and 1840s, when occasionally more than 600 000 pelts were taken, the yield fell off to the point where, in the last two decades of the century, anything over 300 000 was exceptional. This coincided with a drop in seal-oil prices after 1880 as mineral oils became increasingly available. By the 1880s, seal oil accounted for only one-eighth of the value of Newfoundland's total exports and by the end of the century, less than one-tenth.[19]

All this should not obscure the importance of sealing to the Newfoundland economy in the 19th century. For the century as a whole, only salted cod exceeded seal oil in value as an export product.[20] The crisis that hit the Newfoundland fishing industry in the second half of the century, especially after the mid-1880s, would have occurred much earlier if it had not been cushioned by the sealing industry and, to a lesser extent, the Labrador fishery. The population of Newfoundland nearly doubled in the 50 years after 1825, causing serious overcrowding of inshore fishing grounds. The rapid development of the Labrador fishery and the springtime vessel-based hunt for seals during and after the Napoleonic Wars had created alternative employment for the excess population.[21] The decline of these two industries in the last decades of the century put the pressure back on the inshore fishery and prompted efforts to promote a bank fishery, but these could not pick up the slack.

Concentration and Decline

The First World War witnessed the break-up of Newfoundland's entire steel sealing fleet. The Russian government purchased eight steel ships for ice-breaking service in the White Sea, the only sea route not controlled by Russia's enemies.[22] The others were at various times required for wartime service, so by 1916 Bowring Brothers' *Florizel* was the only steel ship able to participate in the hunt.[23] The *Florizel* was wrecked north of Cape Race on 24 February 1918 while on passenger service. Another Bowring steel ship, the *Stephano*, was torpedoed off Nantucket on 8 October 1916, also while on passenger service. Thus the fleet declined from 20 vessels in 1914 to 12, all wooden walls, in 1918. In the same period the number of sealers fell from 3959 to 2056. The displacement of these men was partially offset by the availability of war-related work, especially cutting timber to supply pit-props for British mines.[24]

During the war and in the early postwar period, more sealing firms withdrew from the industry. The withdrawal was so rapid that by 1923 only Bowring Brothers and Job Brothers remained. This phenomenon has traditionally been explained away as the result of over-zealous government regulation of the industry in the wake of two major disasters in 1914 that claimed 250 lives.[25] Others have suggested that postwar profits were too small to permit new ship construction.[26] But conventional wisdom is contradicted by the performance of both Job Brothers and Bowring Brothers, who each added two steel ships to the fleet between 1926 and 1929, financed an aerial spotting service for most of the decade, and introduced new seal-skinning equipment.

Job Brothers and Bowring Brothers may have been in better positions than their competitors to absorb losses. From its beginnings when Benjamin Bowring opened a watchmaker's shop at St. John's in 1811, Bowring Brothers grew into an international corporate empire based on shipping and insurance, with headquarters in Liverpool, England.[27] Although the company history is vague on the subject, some Bowring wooden walls engaged in whaling in Davis Strait in the 1880s and 1890s.[28] Bowring wooden walls were frequently sold or chartered for polar expeditions. The firm's steel ships, when not sealing, carried freight around the world, or else, as part of its Red Cross line, transported passengers between St. John's, Halifax, New

York, and the Caribbean. Bowring Brothers successfully overcame the problem of off-season use of sealing ships.

The older of the two firms, Job Brothers and Company Limited, established in 1730, was the St. John's branch of Job Brothers Limited, a shipping firm also based in Liverpool.[29] As part of a larger organization, Job Brothers, like Bowring Brothers, could sustain losses in its sealing operations that might ruin smaller firms. The similarities between the two extend to Job Brothers' record of finding off-season employment for its sealing ships. The Canadian Department of Marine and Fisheries chartered the wooden wall *Neptune* for Arctic expeditions in 1884 and 1903–04, and for icebreaking work in Northumberland Strait in 1887.[30] More famous was the *Nimrod*, used by Antarctic explorer Sir Ernest Shackleton.[31] Job Brothers also benefited from a lengthy business relationship with the Hudson's Bay Company, which owned 51 per cent of the steel ship SS *Nascopie* (and exercised its option to buy Job's share in 1915). These factors improved the competitive positions of Bowring Brothers and Job Brothers, and at least partially explain why they outlasted their competitors.

Perhaps it is too facile to argue that Bowring Brothers and Job Brothers were simply better businessmen. The sealing industry, like the fishery as a whole, was the victim of general entrepreneurial and government neglect. A.J. Harvey and Company was a prime example of the former. The firm failed to reinvest revenues earned from the sale of its steel ships and never sent another ship to the ice after 1915. Lack of commitment was also evident in government. Beginning in the 1880s, the government had eschewed the fishery, first in an unsuccessful attempt to develop a local manufacturing sector, followed in the 1890s by a program of export-led growth based on "highly marginal land resources."[32] The Bell Island iron ore mine opened in 1895, followed by the Grand Falls newsprint mill in 1905, the Corner Brook newsprint mill in 1923, the Buchans lead, zinc, and copper mine in 1928, and the St. Lawrence fluorspar mine in 1933. Fishery products accounted for 71 per cent of the annual value of Newfoundland's exports in 1920–21, but fell to 37.6 per cent by 1929–30 as the fishery was surpassed in economic importance by the new mining and forest industries. Norway replaced Newfoundland as the world's largest exporter of salt fish in 1909, and from 1920 to 1930 Newfoundland's share of cod landings off North America and Greenland fell from 49 per cent to 41 per cent.[33] Part of the problem was the diversion of capital into servicing the colony's huge national debt, a legacy of over-zealous railway construction and financial contributions to the imperial war effort. But there was also a definite lack of government and business initiative. Of course, the line between government and business in Newfoundland at the time was a thin one: the St. John's mercantile class was well represented in government and exerted considerable, sometimes dominant, influence on it.[34] The plight of the sealing industry in the postwar period must be viewed against this background of negligence towards the fishing industry as a whole.

The Close-Season Debate

In the 1930s the sealing industry felt the effects of the Great Depression. The worst year was 1932, when only four ships went to the ice and for the first time since the industry's formative years, the number of sealers dropped below a thousand. Many speculated that the seals were headed for extinction. The topic was not new. Concern over declining catches had led to the banning of second trips in 1892, which eased the pressure on breeding seals. As early as 1915, one Newfoundlander had predicted that the seals might disappear by mid-century.[35] In 1928 the St. John's correspondent of the *Canadian Fisherman* echoed the feelings of many when he argued that a suspension of the hunt of "several years" duration was necessary if the industry was to survive.[36] Defenders of the hunt countered that catches were low because the ships could not find the "main patch," where the vast majority of the seals were said to congregate. Captain Abram Kean feared that a close season would backfire because during the close season, substitutes might be found for seal products. Barring that, there would be such a demand for seal products afterwards that "the same mad rush would commence, and history would repeat itself."[37] Kean felt that the industry could be sustained at a level of eight or nine ships and 1400 men.[38] At first glance, Kean appears to have been a prophet: the fleet averaged 7.6 ships and 1400.8 men a year during the 1930s. More likely he was just echoing the views of his employer, Bowring Brothers, who were in a better position than most to influence the hunt.

In 1916 the government restricted to two the number of rifles per ship, and in 1931, rifles were banned outright.[39] Eventually the ban was relaxed slightly, each vessel being allowed to carry one rifle.[40] This was a very creditable approach to increasing the seal population because it placed almost the entire focus of the hunt on the seal pups. Since pups have a higher natural mortality rate, they can sustain much greater kill levels than adult seals. These actions, together with the 1892 banning of second trips, were at least partly responsible for stabilizing the seal population in the 1930s.

The stabilizing seal population was the theme of marine biologist J.S. Colman's article on the steamer fleet, published in the *Journal of Animal Ecology* in 1937.[41] Colman maintained that the number of sealers was the key factor in determining the size of the catch and hence in influencing the seal population. In 1915 the number of sealers fell below 3000 for the first time in the 20th century. For most of the 1920s there were fewer than 2000 sealers and in 1932 there were a mere 731. This had an obvious effect on the number of seals taken. During the 44 years from 1871 to 1914 inclusive, the annual catch surpassed 200 000 seals on 32 occasions, or 73 per cent of the time. But in the 22 years from 1915 to 1936 inclusive, the 200 000 level was exceeded on only six occasions, or 27 per cent of the time.

Colman also noticed another interesting trend. Although crews were smaller and catch levels generally lower since 1914, the average catch per man had risen. Between 1863 and 1914, catch per man exceeded 100 seals only once (124 in 1871). From 1915 to 1936 it exceeded 100 on ten occasions. Colman attributed this phenomenon to a partial recovery in the seal herds since the First World War. He suggested that the danger limit for over-harvesting was around four thousand sealers, or almost three times the average annual number of sealers in the 1930s. He therefore concluded that "at the present time and at the present rate of fishing the seals are a little more than holding their own." Although Colman did not have the benefit of a seal census, Canadian marine biologist Dr. David Sergeant echoed Colman's findings a quarter of a century later, stating that the Newfoundland harp-seal catch of the interwar period "may well have been a catch which could have been maintained *in perpetuo* or even increased without detriment to the herds."[42]

Colman's study pointed out, first and foremost, that slack existed within the industry. The increases in catch per man during and after the war should have prompted renewed interest in the hunt among St. John's firms. Job Brothers and Bowring Brothers did expand their operations, but they were the only ones to do so. This lends credence to the picture of a fishing industry suffering from entrepreneurial neglect in all its branches. The second point following from Colman is that the close-season advocates were alarmists. It was unrealistic to expect the emaciated post-1914 fleets to produce catches comparable to prewar levels, yet this is essentially what such advocates did. The surprising thing is that they won so many converts. Perhaps, like the merchants, they had lost faith in the industry. Pessimism was so pervasive that experienced sealers like Captain Bob Bartlett were convinced the industry's days were numbered.[43] In such an atmosphere, the industry could do little but languish.

Auxiliary Schooners

One glimmer of innovation in the interwar period, if it had been fully realized, would have transformed the industry. The Fishermen's Protective Union (FPU) saw the disappearance of the steel ships from the sealing fleet as an opportunity for legislation to prevent the steel ships from ever again participating in the seal hunt.[44] In 1916 the FPU called for new shipbuilding bounties to revive wooden shipbuilding so that schooners, now outfitted with auxiliary diesel engines, could again play an important role. In the process, the economy of Newfoundland, especially in the outports, would be rejuvenated. "Auxiliary schooners" were not new. The Hudson's Bay Company had already proven the utility of auxiliary-powered sailing vessels in northern waters, and they were being used in the Labrador fishery and the Newfoundland bank fishery by 1914.[45] Since the advent of steamers, a few outport vessel owners had continued to send sailing vessels to the seal hunt. The FPU now wanted to expand and improve that participation by making motorized sailing vessels the norm in the industry.

At first, nothing came of the FPU proposal. The FPU did not achieve political power until 1919, when it and the Liberal Reform party formed a coalition government. The union's primary objective then was reorganizing the colony's faltering cod fishery. In 1920 the House of Assembly did adopt the union's legislative program for that fishery, only to repeal it in 1921 after the major fish exporters refused to go along with it.[46] Thereafter, the FPU lapsed into a political torpor and made no attempt to introduce similar legislation to revolutionize the sealing industry. However, in 1927 the union acted on its own and sent the 60-ton "motor schooner" *Young Harp* to the ice. The vessel had been built at Port Union, the site of FPU headquarters, in 1926.[47] Its crew of 27 took 4353 "prime young seals," an average of 161 seals per man, versus 110 seals per man for the steamer fleet that year.[48] Buoyed by this success, the Union Trading Company again sent the *Young Harp* to the ice in 1928, when the crew secured the highest share of the season.[49] Not only did the crew receive half the proceeds from the catch, but the outfitting costs were minuscule compared with steamer costs. Public interest was aroused. In 1929 seven auxiliary schooners participated in the hunt:[50] the *Young Harp*, now owned by the Hudson's Bay Company; *Lone Flyer* and *Arathusa*, owned by Ashbourne and Company Limited, Twillingate; *Humorist*, owned by Monroe Export Company, St. John's; *Dazzle*, owned by W. Wareham and Son, Harbour Buffett; and *Swile* and *FPU*, owned by the Union Trading Company and sailing from Catalina. There were more than half as many auxiliary schooners as there were steamers.

For reasons that are not entirely clear, auxiliary schooners declined after 1930, when six had cleared for the ice. There were only two in 1931, both owned by the Union Trading Company, and for the remainder of the decade the number of auxiliary schooners never rose above three. Existing accounts of the seal hunt reflect an obsession with the steamer fleet, so further research is required on the auxiliary schooner's role. However, one tentative explanation can be offered for its demise. The enthusiasm generated by the early success of the *Young Harp* may not have been entirely justified. Ice conditions in 1928 were ideally suited to auxiliary schooners. Because of mild weather, the ice was loose and scattered, so the seal patches were quite small and easily reached by schooner.[51] These conditions militated against the steamers because it was uneconomical for steamer crews to waste time cleaning up small patches. Conversely, heavy ice was bad for auxiliary schooners, limiting their movements and posing a serious safety hazard. Heavy ice contributed to a poor season for auxiliary schooners in 1929, and in 1931 the auxiliary schooner *Sir William*, owned by the Union Trading Company, was crushed in the ice.[52] It appears that over a period of several years the steamers once again proved their superiority in the ice. Nevertheless, the use of auxiliary schooners represented an innovative approach to solving the ills of the sealing industry at a time when many had prematurely given up on it. In addition, the auxiliary schooner pointed the way to the post–Second World War sealing fleet, in which fully motorized vessels completely displaced steamships, as they did in other branches of the fishery. Finally, the auxiliary schooners posed the first serious threat in decades to industry domination by St. John's firms. For these reasons, the FPU initiative was an important development in the history of Newfoundland sealing.

The Rise of Foreign Competition

While the industry stagnated, the first hints of the future appeared in the form of Canadian and Norwegian interest in the Newfoundland sealing grounds. Sealing had received sporadic attention in Nova Scotia since the early 19th century. In 1828 James Quinlan of Halifax fitted out what was said to have been the first Nova Scotian sealing vessel.[53] It was Cape Breton, however, that came to dominate Nova Scotia's nascent sealing industry. In 1843 Cheticamp and Margaree Harbour sent 22 vessels to the ice, taking almost 10 000 seals.[54] Largely dependent on government bounties, Nova Scotian sealing was fitful, the fleet disappearing "almost immediately" after 1843, only to emerge later. Ninety sealing vessels were reported in the Gulf of St. Lawrence in 1858, including an unspecified number from Halifax and two from La Poile, on the southwest coast of Newfoundland.[55] The vast majority likely came from the Quebec ports of Natashquan and Pointe-aux-Esquimaux.[56] Cunard and Morrow of Halifax sent a single steamer "to the Newfoundland coast," presumably the Front, in 1865 and 1866, but once again this was not a lasting commitment.[57]

In the mid-1880s a new and more committed owner entered the Nova Scotia sealing picture. Captain James A. Farquhar of Halifax began sending ships into the Gulf of St. Lawrence, picking up Newfoundland crewmen at Channel–Port aux Basques.[58] In 1893 Farquhar and Company purchased the SS *Newfoundland* from Allen Brothers of Montreal in a joint venture with A.J. Harvey and Company. The *Newfoundland* concentrated on the gulf, but did occasionally go to the Front. It made its last voyage for Farquhar in 1903, after which it was purchased outright by A.J. Harvey. After sending the *Havana* to the gulf in 1906, Farquhar was absent from the hunt until 1910, when he returned with the *Harlaw*.[59] Absent again in 1911, he returned in 1912 with the *Seal*, which was joined in 1916 by the *Sable I*. Both steamers took on Newfoundland crews and had Newfoundland captains. The *Sable I*, with a crew of 130, cleared for the Front from Pool's Island, while the *Seal* picked up a crew at Channel–Port aux Basques for the gulf hunt. The *Seal* discharged its catch of approximately 3000 seals at North Sydney, where Marden, Orth and Hastings Company of Boston had established a plant to process seal fat into oil.[60] Farquhar ships continued to sail from Newfoundland ports until 1921. The firm, renamed Farquhar Steamship Company, next sent the *Stella Maris* from St. John's in 1925, but the vessel was lost after being crushed in the ice (the crew were saved).

Four years later, the Marden-Wilde Corporation of Boston, in association with J.B. Mitchell of Halifax, followed Farquhar's example by sending the *Vedas* to the ice from Louisbourg. The *Vedas* carried a Newfoundland captain, George Murley, and 94 Newfoundland sealers.[61] The catch was processed at Marden-Wilde's North Sydney plant. This venture was abandoned after 1930. Until the Second World War, Nova Scotia's participation in sealing did not seriously threaten Newfoundland. Nevertheless, patterns were established that were to be repeated later with important consequences for the Newfoundland sealing industry.

Equally significant for the future of the Newfoundland industry was the emergence of Norwegian interest in the region, beginning with the Gulf of St. Lawrence in 1913.[62] In 1916 the Norwegian steamer *Njord* cleared from Louisbourg, and like its Nova Scotian counterparts *Seal* and *Sable I*, carried a Newfoundland crew.[63] Other Norwegian steamers sailed from Sydney.[64]

The gulf catch that year was low and observers suspected that the Norwegians lost badly because they paid their crews wages plus bonuses, instead of shares. Perhaps for that reason the Norwegians did not return the following year. Norwegian interest was partly attributable to depleted stocks at their traditional sealing grounds, the "West Ice" off Jan Mayen. This had already led them into direct competition with the Russians in the White Sea around 1900.[65] The First World War also likely disrupted the White Sea hunt. With the return of peace in 1918, the Norwegians abandoned their experiment in the gulf and concentrated on the White Sea. The Norwegian catch rose from 100 000 seals in 1918 to a peak of 343 000 in 1925. As in the cod fishery, Norway had surpassed Newfoundland.

In 1937, faced with rapidly declining catches in the White Sea, the Norwegians turned once again to North American waters, this time to the Front. A single ship, the *Ora*, was sent out that year. The following year the *Ora* was joined by two more ships, *Polaris* and *Polarbjorn*. Even though the *Ora* was crushed in the ice and sank with its catch of 11 000 seals, the others returned to Norway with 31 000 pelts.[66] The Norwegians quickly realized that the Front could take the place of the White Sea and accordingly sent out eight ships in 1939. These would have outnumbered the Newfoundland fleet of seven steamers except that one of them sunk en route to the Front, and two others were lost while there. The remaining five took roughly 33 000 seals, or one-third the number taken by Newfoundland ships (97 345).[67] Despite the war, five Norwegian sealing ships crossed the north Atlantic in 1940, three of which, *Polaris*, *Polarbjorn*, and *Aretos*, landed their catch in St. John's.[68] The following year a key event occurred when *Polaris* and *Polarbjorn* sailed to the Front from Halifax. This led to the founding of Karlsen Shipping Company Limited, which in turn prompted a revival of Nova Scotia-based sealing.

The Vessel-Based Hunt
1862–1939

Preparations

With a sealing industry dependent on steamships and based in St. John's, there was little of the activity formerly associated with getting sailing vessels ready for the hunt. (A few sailing vessels continued to sail from the outports, especially the more northerly ones, so some preparations along the old lines were necessary, but on nowhere near the same scale.) The steamers were foreign built, mostly in Scottish ports like Aberdeen, Dundee, and Greenock, or else were converted men-of-war no longer required by the Royal Navy.[1] All repairs to the steamers were carried out at the St. John's dry dock, which, when newly renovated in 1884, could accommodate the largest ships of the day. The most important of all supplies was coal, brought by freighter from Cape Breton.[2] At the turn of the century it cost approximately \$11 000 to outfit a sealing vessel, compared with approximately \$25 000 on the eve of the Second World War.[3]

One thing that did not change with steam was the annual stampede for berths, which actually intensified as berths became scarcer. After Christmas and into January, hopeful sealers clamoured around the premises of the local merchants who, as agents for the St. John's firms, were entitled to distribute berths.[4] The successful men then might repair their boots and equipment

10   Sealing vessel in dry dock, St. John's.
Provincial Archives of Newfoundland and Labrador, NA-1682.

while their wives mended their clothing.[5] In the steamer era, gaffs were supplied by the vessel owners.

The sealers were due in St. John's around the last week of February. Before the advent of the railway (1890–93), most of the Conception Bay sealers walked to the capital, and even afterward they often had lengthy treks to the nearest stations. The men pulled their belongings on makeshift sleds with flour-barrel staves serving as runners.[6] Beginning in 1892, sealers from southern Bonavista Bay took the train from Port Blandford, while those from the northern part of the bay walked to the Gambo station.[7] Others, the lucky ones, were picked up by their steamers at places such as Pool's Island near Wesleyville and Seldom-Come-By on Fogo Island. This stemmed from the practice, less frequent after 1914, of having the older wooden walls clear from areas closer to the breeding grounds to increase their chances of success. In 1905, out of a fleet of 22 steamers, 16 cleared from Bonavista Bay and one from Fogo.[8] These ships remained in St. John's until late December or January, then sailed to the ports from which they were to clear. Some of the older wooden walls that remained at St. John's cleared not for the Front, but for the Gulf of St. Lawrence, where ice conditions were usually less dangerous.

Vessels

During the period 1862–1939 there were three main vessel types. The first was the sailing vessel, usually a schooner or a brig. Few remained in the industry beyond the 1880s. There were two types of steam vessels: wooden walls and steel ships, the latter not appearing until 1906. The first wooden walls were patterned after Scottish whalers. The SS *Wolf*, purchased by Walter Grieve and Company in 1863, was 210 net tons and possessed only a 30-nhp (nominal horsepower) engine.[9] They quickly grew more powerful. The SS *Merlin*, acquired by A.M. Mackay of St. John's in 1869 and driven by a 110-nhp engine, was the first wooden wall with an engine strength in excess of 100 nhp.[10] The largest of the wooden walls, at 568 net tons, was the SS *Newfoundland* (renamed SS *Samuel Blandford* in 1916), purchased jointly by James A. Farquhar of Halifax and A.J. Harvey and Company of St. John's in 1893. The most powerful was the 180-nhp SS *Algerine* bought, also in 1893, by Bowring Brothers of St. John's.[11] The average of all 58 known wooden walls surveyed for the period 1863–1924 was 305 net tons with an engine strength of 76 nhp. The engine and funnels of a wooden wall were usually situated towards the stern, behind the mainmast. Unlike a sail-

11 SS *Neptune* heading out the Narrows, St. John's, 9 March 1901.
Maritime Museum of the Atlantic, Halifax, Neg. No. N-7872.

12 SS *Adventure circa* 1912.
Maritime Museum of the Atlantic, Halifax, Neg. No. N-12,966.

ing vessel, which had to be modified for the ice, the wooden wall was built specifically for it. Indeed, a number were used for Arctic and Antarctic expeditions. For example, Admiral Robert Scott purchased the SS *Terra Nova* in 1909 for his ill-fated expedition to the South Pole. The beams, bulkhead, and frame of the wooden wall were made from oak, the sides were sheathed in greenheart, and the bow encased in iron or steel plating.[12] All wooden walls used screw propellors.[13]

Declining catches in the late 19th century prompted the next technological innovation: in 1906 A.J. Harvey and Company sent the first steel vessel to the ice. The SS *Adventure* was 826 net tons and had an engine strength of 213 nhp. The *Adventure* was an instant success, its crew taking 30 000 seals that first season. The competition took notice, and by 1914 there were nine steel vessels, or about half of the entire fleet. The largest and most powerful was Bowring Brothers' SS *Stephano*, which first sailed to the ice in 1912. The *Stephano* was 2144 net tons and boasted a 577-nhp engine. Of the 18 known steel ships surveyed for the period 1906–29, the average was 802 net tons and possessed a 252-nhp engine.

The sealing fleet of the interwar period consisted of a few pre-1914 wooden walls and a handful of steel ships introduced in the late 1920s. Foremost among the new steel ships were Bowring Brothers' *Imogene*, which went to the ice from 1929 to 1940; the *Boethic*, which sailed for Job Brothers from 1926 to 1932 and for Bowring Brothers from 1934 to 1940; and the *Ungava*, which sailed for Job Brothers from 1928 to 1931 and for the SS *Ungava* Company, a Job subsidiary, from 1934 to 1940. The *Boethic*, at 1078 net tons, was the largest of the three, but it was still smaller than the prewar giants *Florizel* and *Stephano*. The *Imogene* distinguished itself with a catch of 55 636 seals in 1933, the biggest ever for a single voyage, but essentially there was no difference between the newer steel ships and their predecessors.

Season Length

The addition of steamers to the Newfoundland sealing fleet led to changing regulations governing departure dates and season length. Because steamers got to the ice so much faster than sailing vessels, sailing-vessel owners complained of unfair competition. In an attempt to restore some fairness, the Newfoundland government passed legislation in 1873 requiring steamers to remain in port until 10 March, but permitting sailing vessels to leave on 5 March.[14] No seals were to be killed before 12 March. The gap between sailing dates for steamers and sailing vessels was lengthened in 1883 to nine days, steamers being enjoined from leaving port before 6:00 a.m. on 10 March, compared with 1 March for sailing vessels.[15] Ships of either type could leave a day earlier if their departure date fell on a Sunday.

In 1887 the government took the first steps to preserve the resource, restricting the season for steamers to the period 12 March–20 April and prohibiting steamers from making second trips after 1 April.[16] This was followed in 1892 by legislation completely banning second trips for steamers and further shortening the season for steamer crews to the period 14 March–20 April.[17] After the banning of second trips, most steamers were back in port by early May. In 1914, wooden walls were allowed to sail any time after 8:00 a.m. on 12 March.[18] Steel ships clearing from St. John's or ports south and west could not depart before 8:00 a.m. on 13 March; steel ships clearing from ports north of St. John's could not leave before 8:00 a.m. on 14 March, since they were already closer to the seals. Crews on both wooden walls and steel ships could not kill any seals before 15 March.

Before 1913, steamer crews competed to be the first out of St. John's harbour. That year, however, the *Boethic* and the *Bonaventure* collided during the race, necessitating expensive repairs and delaying their departure. Thereafter, steamers left in an orderly fashion.[19] Once outside the narrows, though, the "race of the inside cut" began as earnestly as ever.

After 1914, departure dates for Newfoundland sealing ships fluctuated considerably. The Seal Fishery Act of 1916 amended the 1914 legislation by enabling customs officers to clear any wooden ship on 11 March if 12 March were a Sunday.[20] No steel ship was to clear for the ice until 48 hours after the time set for wooden ships. Legislation passed in 1921 set a uniform departure time, but the fleet then consisted entirely of wooden walls.[21] No ship could depart before 8:00 a.m. on 10 March, or 9 March if the tenth fell on a Sunday. The earlier departure date was probably a response to the declining physical condition of the fleet. The trend towards earlier departures continued in 1925, when the departure time was moved up to 8:00 a.m. on 5 March.[22] This was amended slightly in 1926 to permit departure on 6 March when 5 March was a Sunday.[23] Also, seals could be taken any time after 13 March, two days earlier than previously. In 1939 and 1940 only, sealing ships were allowed to clear on 3 March, killing to begin on 10 March. This was in response to the Norwegians, who began sealing well before 13 March.

Navigation

While steam gave man a better chance in the ice, it by no means gave him dominion over it. New ice-breaking techniques were developed but, surprisingly, many of the old ones were retained. Thus there evolved a hybrid system borrowing from sailing ships and incorporating new techniques peculiar to steamships. Ice-saws, ice-chisels, axes, hawsers, and stabber poles remained standard equipment. The wooden walls also possessed rams, called "sheer poles," by the 1870s.[24] (Perhaps the most curious carry-over was the use of sails. This might have been expected on the early wooden walls, which were bark or barquentine rigged.[25] In the British whaling industry, steamers were outfitted with sails because whalers felt that engine noise scared off the whales once the ships were on the breeding grounds.[26] In the seal hunt, general opinion held that not only noise, but also smoke from the steamer's engines warned bedlamers and adult seals of the hunters' approach.[27])

The steamer's great advantages were its extra strength, manoeuvrability, and speed (eight to ten knots in open water).[28] Unlike sailing vessels, steamers were able to move astern as well as forward. In heavy ice the steamer reversed and then ran forward at the ice. This action, known as "butting," was repeated until the ice gave way.[29] The steamer's bow was steeply raked so that the ship rode up onto the ice and crushed it under the ship's weight. Hence the superiority of steel ships over the lighter wooden walls. The bow of the SS *Bear* projected a full 20 feet from the stem.[30] The inside of the bow, known as the "fortification," was strengthened with beams, timbers, and diagonals.

Because steamer crews were so large, another new ice-breaking technique was for the men to run from side to side on the deck, creating a rolling motion that relieved pressure on the sides.[31] Gunpowder was sometimes used to break up heavy ice in front of the ship or to remove pressure ridges in the ship's path. An account of the 1874 seal hunt makes no mention of gunpowder; it probably did not come into vogue until the early 20th century. Powder was poured into a tin called a "bomb," which had a waterproof fuse at one end.[32] The bombs were thrust beneath the ice by means of stabber poles. As soon as the fuses were lit, the men scattered to escape the shower of exploding ice that would follow.

Often, not one but several methods were employed at the same time to free a steamer, methods made possible again by the larger crew. This differed sharply from a sailing vessel where the entire crew was usually assigned to a single task.[33]

13 Sealing fleet in St. John's harbour, 1880.
National Archives of Canada, C-73068.

14 Cutting a channel in the ice, St. John's harbour, 1880.
National Archives of Canada, C-10458.

15 Sealers preparing to pull a ship through the ice, *circa* 1912.
Maritime Museum of the Atlantic, Halifax, Neg. No. N-12,964.

16 Sealers with a hawser, ready to pull a ship through the ice, *circa* 1912.
Maritime Museum of the Atlantic, Halifax, Neg. No. N-12,963.

17 Sealers using stabber poles to free the SS *Florizel, circa* 1914.
Provincial Archives of Newfoundland and Labrador, A9-15, Holloway Studio.

18 Sealers using a hawser and stabber poles to free a ship, *circa* 1912.
Maritime Museum of the Atlantic, Halifax, Neg. No. N-12,962.

19 Wooden walls following in the wake of the van ship, *circa* 1933.
Major William Howe Greene, *The Wooden Walls Among the Ice Floes: Telling the Romance of the New-foundland Seal Fishery* (London: Hutchinson, 1933), facing p. 141; National Archives of Canada, C-18219.

Instantly the whole [steam] ship's company wakes to life and ac-tion — there is no need for any one to ask what has happened. The sleepers come rushing up from below. Long poles are hauled out by some and thrown onto the ice to men already there. Others jump over the bulwarks armed with great ice-chisels, each one of these a foot across the blade and fitted with heavy handles five feet long. There are men with cross-handled saws of about the same length as the chisels; and all these men get at once to work cutting the ice all around the ship into sheets, while others with axes chop round the bow and sides amid a shower of flying ice-splinters. All know what to do and there will be no confusion, while everywhere round the ship will be seen the hefty crew with each man working his hard-est, knowing how much depends on it.

The Captain now orders out a heavy hawser; and almost before his words are spoken, one coil is following so quickly after the other that in a few tens of seconds the end of it is some 200 feet ahead, so great is the impetus of the racing men. Then half a hundred stalwart fishermen tail on to it as if for a mighty tug-of-war; pulling the bow down and striving to clear her by breaking down the ice that lies in front and to the sides of her, every man straining every nerve and muscle on the rope. A bomb is placed ahead and fired with hardly a warning cry — but with little visible effect, perhaps luckily for the heedless men.

On board a close-packed crowd runs at the word of command from one side to the other, leaning far out over each bulwark. It seems a great game of "Tag" with their shouts of "Over-r-r-r" and "Back again", as they strive by their moving weight, to roll her clear. Like a lot of children they push and shove, and lean on each other as they reach the rails at the end of each rush; yet though they shout and laugh, all is in deadly earnest. For it is by this crowding of their weight together that they effect a greater rolling leverage on the hull.

As there is no move yet — the direction of haulage is reversed. Back to the stern the hawser goes from beyond the bow, and again comes the dragging of its heavy coils at an express speed. Then, with some of the crew still rolling her, with others loosening the ice that has been already cut, pushing great pieces of it under the Floe with the risk of a ducking unheeded — with engines reversed and boilers at their fullest power, the whole of the men left free will haul and drag on the hawser's end — panting, sweating, and pull-ing.[34]

Now it was acceptable for other ships to follow in the wake of the steamer that had successfully forged a channel through the ice. (The first ship in the line was called the "van ship.") In the days of the sailing vessels, when the emphasis was on human as opposed to mechanical effort, it was unaccept-able to trail another vessel.[35]

The ice was not always hostile. Steamers often took refuge on the leeward side of icebergs where they found shelter from the wind.[36] Sometimes a row of icebergs caused the ice to pile up on the windward side, creating open water on the leeward through which the steamers navigated freely. At night, steamers were routinely anchored in the ice field by means of large grapnels called "ice claws," or cables wrapped around ice pinnacles.[37]

Crews

With the advent of steamers, owners became more particular than ever in selecting crews. Because the total number of berths was declining, only the very best men could expect to get berths on steamers. The others, usually the older men, were consigned to sailing vessels.[38] Everybody wanted to serve on steamers because of the greater likelihood of taking seals and because steamers were perceived to be safer and more comfortable.

Steamer crews started big and grew bigger. The crews of the first wooden walls, SS *Bloodhound* and SS *Wolf*, at 100 and 110 men respectively, were twice the size of the crew of an average sailing vessel.[39] In 1873 five steamers had crews of over 200 men; nine years later the average crew size for the steamer fleet as a whole surpassed the 200 level. Even then there were larger crews. The SS *Thetis*, for example, carried 319 men in 1882. These levels were reached partly on account of the steamers' size, but primarily at the expense of the men's comfort. In an effort to prevent further overcrowding, the government passed legislation in 1898 setting a maximum limit of 270 men per steamer.[40] From 1914 to 1940 the average crew size of a Newfoundland sealing steamer normally hovered around 180 men. The chief exception was the six-year period 1921–26 when the average ranged between 137 and 151.

The crews continued to be divided into watches and punt crews. In 1874 the 273 men of the SS *Bear*, not all of them sealers, were divided into three watches.[41] After legislation limited crew size in 1898, three or four watches were the norm.[42] Each watch was subdivided into ice parties, each commanded by an ice master or quartermaster who answered to the master watch.[43] Selecting steamer watches was more arbitrary than selecting watches on sailing vessels because of the sheer numbers involved. The master watches drafted men from the ship's list. This was not as sensitive to kinship and community ties, although men could and did change watches with their master watch's permission.[44]

On the *Bear* each punt crew consisted of four men: one "bow" gunner, an "after" gunner, and two oarsmen, who presumably also functioned as "dogs."[45] Only a portion of the sealers on steamers were selected for punt duty, which was not the case on sailing vessels, where all sealers saw punt duty. The SS *Bear*, for example, had 25 punts, meaning that less than one-third of the sealers were eligible for punt duty. This was the result of a reduction in the importance of the beater and adult seal hunts after the introduction of steamers. Because steamers got to the ice so much faster than sailing vessels, they were more likely to encounter whitecoats. In addition, the financial returns from the beater and adult seal hunts frequently were too small to offset the steamers' high operating costs.[46] It is true that until the banning of second trips, some steamers made second and third voyages. However, they usually did so with significantly smaller crews.[47] Still, if a steamer did miss the whitecoats, it could always fall back on beaters and adult seals to reduce losses or perhaps turn a profit. And, even if a sizable number of harps were taken, captains were reluctant to return to port if they still had space for more pelts. Thus most steam voyages resulted in a go at the "old fat."

Sealers' shipboard duties do not appear to have been diminished by the advent of steam since plenty of new work was associated with coal. The "ash-cat" gang used a windlass to haul ashes up from the boiler room in steel buckets that they emptied over the side.[48] The coal bunkers had to be replenished continually, either from coal pounds on deck or from the holds below deck. Also, coal used as ballast had to be removed (usually thrown overboard) as pelts were brought aboard.[49]

Once the ship's water tank ran empty, as it invariably did, men had to go onto the ice to get "pinnacle ice" for drinking water (the captain and officers had their own water on ship). Pinnacle ice was fresh-water ice obtained from either glacial icebergs or frozen rainwater.[50] The men hacked off pieces of ice with axes and melted it in a tank located at the base of the vessel's funnel and heated by steam piped from the boilers.[51]

While en route to the seals, the sealers constructed "side sticks" or "baulks," rough ladders used in leaving or boarding the ship.[52] Their equipment and clothing required constant attention. They sharpened their sculping knives (on a grindstone located near the bridge), adjusted gaffs, spliced tow ropes, and patched their mittens and boots.

In addition to the sealers themselves, steamers carried a number of solely shipboard personnel. One of the most vital was the scunner or "barrelman" who, from his position in a barrel atop the foremast, spied a course through the ice and also kept track of the men once the hunt began.[53] He shouted directions to the bridge-master below, who in turn relayed them to the wheelman (also "steersman"). Three or four wheelmen took turns at the wheel. The captain or some other person such as a master watch occupied another barrel, this one atop the mainmast, from which vantage point he searched for seals. The bosun was responsible for much of the ship's equipment, ranging from gaffs, ropes, and compasses to lamps and candles.[54] He was also entrusted with keeping the deck clear, a hopeless task once the ship was "in the fat." With steam engines came the need for engineers and firemen.

Steamers also carried cooks and stewards, the latter exclusively at the service of the captains and officers.

Before leaving this section, a word must be said about the home communities of the sealing captains and the men. Before 1914, captains and crews came from northern bays, particularly northern Bonavista Bay. In 1890, for example, half the sealing captains were from northern Bonavista Bay.[55] Only detailed examination of a number of crew lists will reveal what proportion of the sealers came from that area; however, we do know that half the crew of the SS *Newfoundland* in 1914 were Bonavista Bay men.[56] The homes of sealers after 1914 have not been studied in the same detail as for the prewar period, but newspaper accounts for the 1920s and 1930s indicate that Bonavista Bay was still an important source, perhaps the dominant one.

Equipment

Seal hunters' basic equipment changed little after the introduction of steamships. The gaff, tow rope, and sculping knife were still the chief tools. Although the gaff seems not to have undergone any radical changes, the tow rope was slightly modified in that the tapered end was replaced by an iron or steel hook.[57] Tow ropes were cut while the steamers were en route to the ice.[58] The sculping knife evolved from the clasp knife mentioned in 1840 to a sheath knife, the sheath being a homemade affair formed by two pieces of wood bound together by twine.[59] The knife, essentially the same one used to split codfish, had a wooden handle and a thin, curved blade five to six inches long.[60] In addition to sharpening his knife on a grindstone located on the deck, each man carried a small sharpening steel for use on the ice.[61]

The sealing punt was joined by the dory for use in loose ice. The dory, of New England origin, was a flat-bottomed boat with flared sides and pointed bow and stern.[62] Newfoundlanders would have learned of it from Nova Scotian and New England bank fishermen, or else from the St. Pierre and Miquelon French, who imported New England dories for service in their own fishery throughout the 19th century.[63] The dory was widely adopted by Newfoundland bank fishermen in the 1870s. Each punt or dory was numbered for identification purposes.[64] Bank fishermen, who fished one or two to a dory, would have been appalled by the overcrowding in the sealing boats — one account tells of seven men aboard a single boat.[65] Winchester rifles and .44-calibre cartridges replaced long-toms for shooting beater and adult seals.[66] Soft-nosed dumdum bullets were popular. At least one new item was added to the outfit of the punt and dory crews. This was the "seal dog," an iron hook attached to a rope or chain and used to haul large seals to the boat or vessel.[67]

Some of the better descriptions of sealers' clothing come from this period:

The outfit of the sealers is of the simplest description. Sealskin boots reaching to the knee, having a thick leather sole well nailed, to enable them to walk over the ice, protect the feet; coarse canvas jackets, often showing the industry of a wife or mother in the number of patches which adorn them, are worn over warm woollen shirts and other inner clothing; sealskin caps and tweed or moleskin trousers, with thick woollen mits, complete the costume, which is more picturesque than handsome.[68]

Another source from the period also mentioned the sealers' woollen clothing.[69] However, a commission of enquiry into the 1914 seal hunt found that many sealers wore fleece-lined cotton clothes. Such material was "wholly unsuitable for men who have to take violent exercise as, unlike flannel, it fails to absorb perspiration, becomes wet and chills the body."[70] The commissioners recommended that its use be prohibited in future hunts. Sealers also wore coloured-glass goggles to guard against ice blindness.[71]

The Whitecoat Hunt

The deployment of steamship crews allowed more efficient and widespread hunting than was possible for sailing-vessel crews. Because steamships were more mobile and their crews so much larger, a far greater area could be covered. Ice conditions were not so crucial to the deployment pattern. Instead, the shape of the patch was the prime consideration.[72]

Usually the watches gathered on deck at sunrise and went out at the captain's command.[73] The first watch was dropped off around 5:00 a.m., the last by 7:00 a.m.[74] However, under ideal conditions (a clear night sky and a full moon), the men might be on the ice well before the normal hour.[75] Then they would be called back by the ship's whistle for breakfast around 6:00 a.m., after which they would return immediately to the ice. Only a skeleton crew remained on the ship: usually the captain, scunner, engineers, cooks, and a few men to winch pelts aboard. Once on the ice, each watch marched to the seals and fanned out in a circle to work its allotted area. The master watches had compasses so that if the weather turned bad, they would have a general idea where they might find the ship. The men broke up into units under the master watches and the ice masters. Within these smaller units they worked in pairs as they had when operating from sailing vessels.

The first task of the master watches and ice masters was to select large, strong ice pans to serve as collecting areas for the pelts. A "pan flag" bearing the ship's insignia was fixed into the chosen pan.[76] Instead of towing their pelts to the ship, the sealers towed them to the collecting pans. This practice, known as "panning," was the most important on-ice innovation of the steamer operation. Pelts were only towed directly to the ship when the seals were too few to warrant sending out watches or when the ship was jammed in the ice. Panning did not make the men's work any easier, but by reducing towing distances it enabled them to spend more time killing seals.

Panning was extremely wasteful and played an important role in the over-harvesting of the seal population. Ideally a steamer picked up all the panned pelts when it returned for its watches or it completed the pickups at night, the men having marked the pans with kerosene torches. However, weather and ice conditions sometimes made this impossible. Also, a few captains preferred to leave the pelts on the ice overnight to cool, then picked them up after the watches had been dropped off in the morning.[77] Occasionally these pelts were lost, or else stolen. Sealers, with the full blessing of their captains, regularly stole pelts belonging to other ships if there were no witnesses, especially at night or in a fog. They rationalized their behaviour by assuming that any unclaimed pelts had been lost and therefore were fair game.[78]

The actual method of killing the seals, with a gaff, was no different from the method used in the sailing operation. The seals

> *are killed by blows on the nose with the gaff, and are then scalped [sic], by drawing a line with the knife through the skin and blubber from chin to tail, and skinning until the ribs on the left side are reached. The knife is then stuck in the heart, to make a hole [in the carcass] through which a finger can be thrust to grasp a rib, and the carcase is held in that way till the pelt is removed. The scudders, or hinder flippers, are cut off, and when "panning," one of the foremost paws is taken out to make a hole through which to pass the slings for hoisting on board; but when towed to the ship both [fore flippers] are left in to be eaten afterwards.[79]*

For towing, the sealer inserted the hook at the end of the tow rope into the eyeholes of the pelts, put the rope over his shoulder, and hauled the tow to the pan, hair side to the ice. This was probably done to save the fat from

20 Wheelmen on a wooden wall, *circa* 1933.
Major William Howe Greene, *The Wooden Walls Among the Ice Floes: Telling the Romance of the Newfoundland Seal Fishery* (London: Hutchinson, 1933), facing p. 140; National Archives of Canada, C-18218.

21 John Abbott, sealer, *circa* 1900.
Provincial Archives of Newfoundland and Labrador, A3-3, Holloway Studio.

22 Four sealers, *circa* 1933.
Major William Howe Greene, *The Wooden Walls Among the Ice Floes: Telling the Romance of the Newfoundland Seal Fishery* (London: Hutchinson, 1933), facing p. 37.

23 Sealers with dories, *circa* 1912.
Maritime Museum of the Atlantic, Halifax, Neg. No. N-12,959.

24 A watch of sealers, *circa* 1912.
Maritime Museum of the Atlantic, Halifax, Neg. No. N-12,958.

25 Watches heading for seals, *circa* 1914. SS *Adventure* in the background.
National Archives of Canada, PA-121934.

26 Pan of seal pelts marked by the SS *Florizel*'s flag, *circa* 1914.
Provincial Archives of Newfoundland and Labrador, A18-22, Holloway Studio.

27 Clubbing seals, 1881.
National Archives of Canada, C-76630.

28 Towing pelts back to the ship.
Provincial Archives of Newfoundland and Labrador, B1-189, Holloway Studio.

29 Sealers preparing to winch pelts aboard ship, *circa* 1912.
Maritime Museum of the Atlantic, Halifax, Neg. No. N-12,956.

30 Deck scene.
Provincial Archives of Newfoundland and Labrador, NA-1710.

31 Men of the SS *Beothic* cutting ice, *circa* 1912. Side sticks, the sealers' ladders, hang well above the ice.
Maritime Museum of the Atlantic, Halifax, Neg. No. N-12,960.

being chewed up by rough ice and also to make towing easier by reducing friction.[80]

The men's day did not end until the steamer returned for them and their pelts, frequently after dark. The time and place of pickup were prearranged by the captain and master watches.[81] The men boarded the steamer via the side sticks, always on the windward side, where the ice was denser than on the leeward.[82]

The captain picked up his ship's pelts in the same order in which they had been panned; that is, starting with those panned by the first watch. This gave all the pelts the maximum time possible to cool. Three or four men threaded the winch rope, or whipline, through the flipper holes, with 10 to 12 pelts being winched aboard at a time.[83] Loading frequently went on well into the night. Night work was possible because pans not cleaned during the day were marked by kerosene torches.[84] Most of the pelts remained on deck for a night or two, depending on the temperature, to reduce the tendency of the fat to run to oil in the holds. Some were put below immediately, otherwise the deck would have been hopelessly crowded.[85]

When the pelts were ready to be stored below, a strict count was kept on tally sticks (or boards) as they were passed down the hatchway. This task was usually entrusted to the senior master watch.[86] The tallyman cut a notch in the stick — hung from a nail in the mast — for every five pelts, and a groove when the number reached 21. The pelts were stowed in wooden pounds in the holds, laid fat to fat and hair to hair, with crushed sea ice called "sish" or "salt" between each layer. The ice was hacked off alongside the ship and passed by hand onto the deck near the forward hatch, where two or three men chopped it up into smaller pieces.[87] It was then shovelled into flat osier baskets that were passed down to the men in the hold. If, during the course of the voyage, the pounds in the hold became full, new pounds were constructed on deck.[88] If these in turn were filled, the men gave up their bunks to the ubiquitous pelts, after which they slept on top of the pelts, in the coal bunkers, or on the deck itself.[89]

The Beater
and Adult Seal Hunts

The role of punt crews in the beater and adult seal hunts of the steamer era is obscure. The few available sources, most from the early 20th century, indicate that beaters and adults were hunted not from punts, but on foot. The

32 Hunting adult seals, 1880.
National Archives of Canada, C-72982.

move away from punts probably began after the banning of second trips in 1892, since the sealing ships were now less likely to be at sea long enough to encounter loose ice and open water of the type that had necessitated extensive use of punts. Also, this phase had become less economical with the advent of steamers. Nevertheless, punts remained necessary whenever open water lay between the ship and the seals.

The stealth that characterized the beater and adult seal hunts began even earlier under steam. As the steamer neared the seals, the engine was cut back so the smoke and noise would not scare the prey off.[90] In preparation for their work, the dogs stuffed their sealskin or canvas "nunch bags" (also called "nunny bags") with cartridges, as many as 250 per bag.[91] They also put pieces of hard bread in the bags for snacks or emergency rations. The dogs carried the nunch bags in their hands, as opposed to around their necks, so they could let go immediately if they fell in. Russian seal gunners of the late 19th century wore white outer clothing for camouflage, but there is no indication that Newfoundland sealers did so.[92] Instead, they did their best to remain hidden as they crept up on the herd.

When the gunner killed a seal it was the dog's responsibility to sculp it. He usually did this on the ice and then towed the pelt to the nearest pan.[93] Sometimes, if the steamer was nearby, the dog towed the seal to it, where it was winched aboard and sculped on deck.[94] The gaff was useless for hauling a 900-pound hood seal, so the dogs used seal dogs, ropes, or chains with iron hooks. It took several men to haul a round (unsculped) adult hood.

At times, when the steamers became stuck in the ice and there were no seals nearby, the sealers resorted to "swatching."[95] The gunners and their dogs travelled great distances from the ships and waited for seals to appear in open water at the edge of the ice. Because they were in prime condition, seals shot in open water would float for a while, sometimes long enough for dogs to hook them with gaffs and pull them onto the ice. For their comfort the men brought boards to sit on, and built makeshift ice-houses in attempts to stave off the cold. They stayed out on the ice until the captain signalled for them to return. The signals were an ensign flying from the masthead and, in foggy weather, a blast of the ship's steam whistle.

When the gunner sculped a seal, he removed the tip of its tail and threw it into his nunch bag. At the end of the day, the gunners and dogs assembled in the main cabin where, under the captain's scrutiny, each dog emptied his bag and the tails were tallied so the captain knew how many seals had been panned, thus saving him from looking for pelts that did not exist.[96] Captain Abram Kean claimed to have invented this method of counting beater and adult seal pelts.

The Aerial
Spotting Service

A little-known aspect of the vessel-based hunt in the 20th century is the aerial spotting service. The success of aerial reconnaissance missions during the First World War portended untold applications to private industry. The sealing industry, hitherto dependent on the knowledge and instincts of the captains to locate seals, stood to gain obvious advantage. The owners needed some prompting in the form of extremely low catches in the immediate postwar period. In 1920 the catch of 33 985 seals was lower than for any season since 1886 and before the year was out, the owners were negotiating with Major K.E. Clayton-Kennedy of the Aircraft Manufacturing Company Limited. Talks fell through early in 1921, but the owners quickly found a new contractor in Major F. Sidney Cotton, an Australian aviator and First World War veteran who, along with mechanics and other Aircraft Manufacturing personnel, had been at Botwood since before Christmas awaiting word from Clayton-Kennedy.[97]

Cotton made a few flights over the ice in March 1921 but found no seals. The number of flights was restricted by mechanical problems, including frozen radiators and engines. Cotton also had trouble finding the proper skis to enable the plane to land on the ice at Botwood.[98] At the request of the owners, the Newfoundland government agreed to bear half the cost of allowing Cotton to retain his mechanics at Botwood until the end of June so he could get ready for the next season. On 16 March 1922 Cotton and Newfoundlander Victor Sydney Bennett in Martinsyde and Westland biplanes respectively flew over the ice. Cotton spotted a large patch, but he and the owners could not come to terms.[99]

In April 1922 Cotton suggested that the owners experiment next season with operating an airplane from the deck of one of the steamers.[100] The feasibility of this arrangement had been demonstrated by Sir Ernest Shackleton, who had taken a Baby Avro biplane on the steamer *Quest* during his Antarctic expedition in 1921. The owners accepted Cotton's advice and purchased Shackleton's two-seater Baby Avro from A.V. Roe Company Limited of England.[101] Cotton himself was unavailable in 1923 because he was preoccupied with operating the Newfoundland and Labrador air-mail service, which his Aerial Survey Company (Newfoundland) Limited had inaugurated in 1921, so he arranged for an English pilot to fly the spotting plane.[102] That year the *Neptune*, a Job Brothers vessel, carried the Baby Avro to the icefields aboard a specially constructed platform located aft. According to Canadian aviation historian Frank Ellis, the captain of the *Neptune*, George Barbour, was so hostile to the plane that he did not let the pilot make a single flight.[103] The pilot allegedly ended up killing seals with the men to earn his pay. As colourful as this story is, it does not hold up under scrutiny. The 1923 season was marked by strong winds and many snowstorms. Sealing captains agreed that it was one of the worst they had ever experienced. And, as Barbour himself informed an *Evening Telegram* reporter, "even if weather conditions proved favourable, it would have been

33 H. Stannard, an engineer (left), and Major Sidney Cotton at Botwood, 1922.
National Archives of Canada, PA-140793.

34 Westland biplane before a seal-spotting flight, Botwood, 1921.
National Archives of Canada, PA-74266, John Maunder Collection.

35 Baby Avro on a sealing ship.
National Aviation Museum, Ottawa, Neg. No. 2193.

36 SS *Nascopie* with the Baby Avro aft.
National Aviation Museum, Ottawa, Neg. No. 2285.

37 Baby Avro on the ice prior to takeoff.
National Aviation Museum, Ottawa, Neg. No. 2217.

38 Left to right: Roy Grandy, H.E. Wallis (mechanic), and Jabez Winsor on the SS *Eagle*, March 1924.
National Archives of Canada, PA-57841, John Maunder Collection.

39 Avro Avian, a two-seater biplane, *circa* 1928.
National Aviation Museum, Ottawa, Neg. No. 2849.

40 Avro Avian landing alongside the SS *Thetis*, *circa* 1928.
National Aviation Museum, Ottawa, Neg. No. 2276.

impossible to have used the 'plane, as ice conditions would not have permitted a safe take off."[104]

The airplane fared better in 1924, when it was carried on Bowring Brothers' *Eagle*, captained by Edward Bishop. The pilot that year was Captain Roy S. Grandy, a Newfoundlander who had served with the Royal Flying Corps during the First World War. After a period of several days during which few seals were sighted from the ships, Grandy persuaded Bishop to let him make a flight. The Baby Avro, which could be fitted with skis, floats, or wheels as the occasion demanded, required a runway of only about 150 feet.[105] The sealers cleared a strip on the ice and Grandy took off with master watch Jabez Winsor in the passenger seat. They soon spotted a huge patch of seals that the *Eagle* was able to reach. Later in the season, Grandy attempted a second flight, this time from open water, but failed to take off because the airplane developed a bad vibration.

The success of Grandy's first flight in 1924 ensured that, for the time being, the spotting service would continue to play a role in the hunt. Grandy was succeeded in 1925 by C.S. (Jack) Caldwell of Laurentide Air Services Limited of Quebec. Caldwell convinced the owners after the 1927 season that it would be more advantageous to operate the spotting service from land, so the owners acquired an Avro Avian, a two-seater biplane with a range of 500 miles. Before the end of 1927, runways and temporary fuel depots each containing 600 gallons of gasoline were established at Fogo, St. Anthony, Port Saunders, Bonne Bay, and in Labrador, Battle Harbour.[106] The position of the seals was relayed to the fleet via wireless stations on land.

The decision to revert to land bases probably resulted partly from safety considerations. The seal reconnaissance flights constituted "probably the most hazardous phase of flying ever undertaken in Canada."[107] If the ice were to break up while the pilot was aloft, he would be without a landing strip. Or, if a sudden fog or storm arose, he might lose visual contact with the fleet. And for a runway, he had the heaving north Atlantic.

The system that Caldwell pioneered in 1928 had more going for it than just safety considerations. It also allowed the most comprehensive observation ever of the seal herds because flights were more numerous and covered a much wider area. Unfortunately, Caldwell's successor, Alex Harvey of the Ontario Provincial Air Service, wrecked the Avian when he crashed at St. Anthony early in the 1930 season.[108] Had the spotting service continued into the 1930s, seal population surveys might have been conducted. If they had, sealing might have been better regulated in the early postwar period. By

1930, however, the Great Depression had struck and the spotting service saw only sporadic revivals during the next 16 years.

Reaction to the use of spotting planes was mixed. Although some sealing captains were supportive, the most successful of them all, Abram Kean, was not. As he put it in 1935, "I challenge successful contradiction that it was ever worth five dollars to us."[109] Any hostility among the sealers was tempered, initially at least, by awe, for most had never seen an airplane before. Still, that awe did not blunt their sense of humour. The large patch of seals that Cotton claimed to have discovered on 16 March 1922 was thereafter referred to as "The Cotton Patch."[110] Many Newfoundlanders felt that the climate would make springtime flying impractical, but this fear was quickly put to rest.[111] Others believed it would be impossible to see the seals from the air because airplanes were too fast. One of the few objections with any credibility was that even if seals were sighted, it would not always be possible for the ships to reach them. To this the airplane's defenders replied that

> *the sealing master will be in possession of definite information, and at the first slackening [of the ice] can proceed direct to the patch, without any loss of time. How often has it occurred that steamers have passed close by the seals without discovering them, and wasted considerable time searching areas that the aeroplane could have previously pronounced as fruitful or otherwise?[112]*

The impact of the airplane on catch levels in the 1920s cannot be measured accurately. However, in general it appears to have been insignificant. In the five-year period 1923–27, the ship carrying the airplane was the highliner (the ship with the largest catch of seals) three times — 1923, 1926, 1927 — but the airplane was not used in 1923 and in 1927 Caldwell made only one flight and failed to spot any seals. While the average catch per man was high during the 1920s, surpassing 100 seals on five occasions, this was due to a decline in the number of sealers and a recovery of the seal population. In the 1930s, when the airplane was rarely used, average catch per man still exceeded 100 five times. There are two main reasons for the airplane's ineffectiveness. First, the limited ability of the wooden walls to penetrate the ice seriously compromised the airplane's usefulness. Second, while the airplane was based on the sealing ships, ice conditions sometimes made it impossible for it to take off, which rendered it completely useless. Details of the owners' contract with Cotton in 1922 make it possible to quantify the airplane's contribution. The owners determined that 15.9 per cent of that year's catch (20 000 out of 126 031) was attributable to the spotting service.[113] Cotton received ten cents per seal, or $2000, half from the owners and half from the Newfoundland government. The cost-sharing agreement may have collapsed in 1930 as the government approached bankruptcy.

In the support of aerial spotting, Job Brothers and Bowring Brothers demonstrated a continued awareness, apparently not shared by other Newfoundland businessmen, of the potential for profit in the sealing industry. Nevertheless, the spotting service was worthless without a corresponding improvement in the sealing fleet. The addition of three steel ships between 1926 and 1929, accompanied after 1927 by a ambitious land-based spotting service, suggests that the lesson had sunk in. The crash of the spotting plane in 1930 and the onset of the Great Depression virtually ended a scheme that had held out the promise of a more profitable and more sophisticated industry.

Health and Safety

With the advent of steamers and subsequent development of panning, conditions both on the ships and on the ice deteriorated. On the ice, panning had several important consequences. Because the captains tried to cover as much territory as possible in their faster, more mobile ships, the men remained on the ice for longer periods of time than in the days of sail. This also meant that they were left on the ice at much greater distances from their ships, sometimes as much as 10 to 12 miles away. FPU president William Coaker observed in 1914 that one of the steel ships, SS *Nascopie*, did not get all of its men on board until long after dark.[114] They were dropped onto the ice between 5:00 a.m. and 7:00 a.m. and, given a late March sunset of around 6:00 p.m., they would have been on the ice for roughly 12 hours.[115]

In 1890, 540 men belonging to four steamers became separated from their ships in a blizzard and probably would have perished had they not been saved by another ship, the SS *Kite*. On 21 March 1898, 154 men of the SS *Greenland* were stranded on the ice when a storm came up. The wind was so strong that the ship was jammed, in fact almost capsized, while a stretch of water several miles wide opened up between the sealers and their ship.[116] Forty-eight men drowned or froze to death — many of the bodies were never found — and 65 others were severely frostbitten. Tragically, the men had been put onto the ice in the first place because the sealers of another ship, SS *Aurora*, captained by Abram Kean, had allegedly stolen their pans.[117] Otherwise, the *Greenland*'s men would have been picking up their pans, not

41 A steward and cook outside the galley of a wooden wall, *circa* 1933.
Major William Howe Greene, *The Wooden Walls Among the Ice Floes: Telling the Romance of the Newfoundland Seal Fishery* (London: Hutchinson, 1933), facing p. 81.

hunting. An even greater tragedy happened in 1914, when 132 men of the SS *Newfoundland* were lost on the ice for two days during a snowstorm. Of that number, 77 died and the others were badly frostbitten.

Steamers brought entirely new hazards. In 1874 the boilers of the SS *Tigress* exploded, killing 25 men.[118] In January 1882 the SS *Lion*, sailing from St. John's to Trinity to clear for the ice, disappeared in fine weather and was never seen again. It, too, may have been the victim of a boiler explosion, or perhaps a spark ignited the ship's gunpowder supply. Newfoundland sealers "seem to have no prescience of perils. At the beginning I got used to seeing them sit on powder cans and calmly smoke their pipes. Later I observed them filling bombs, still smoking."[119]

Safety got worse, not better, after the introduction of steel steamers. To compete, captains of the wooden walls took unnecessary risks, pushing their ships into ice from which they sometimes could not escape. Ten wooden walls were claimed by the ice from 1907 to 1912. The *Newfoundland* disaster of 1914 was compounded by the loss of the SS *Southern Cross* the same year. The *Southern Cross*, which had been sealing in the Gulf of St. Lawrence, was endeavouring to be the first vessel back from the ice. Its captain, George Clark, tried to sail through a furious storm off the south coast of Newfoundland, with the result that the ship and its entire crew of 173 men were lost east of Cape Pine.[120]

If it can be imagined, living conditions on steamers were worse than those on sailing vessels. The same steamers that had carried 30 to 40 men as their complement for whaling carried up to 200 for sealing.[121] The sealers slept fully dressed in wooden berths located in the forecastle and in the hold ("'tweendecks"), two to three men per berth directly over the pounds where the pelts were stored.[122] The sealer's mattress, known as a "donkey's breakfast," consisted of a cloth sack stuffed with straw or wood chips, and each man had to supply the stuffing himself. The 'tweendecks ceiling was so low that only short men could stand erect.[123] Even these cramped quarters were given over to seal pelts towards the end of most voyages. The captain, master watches, chief engineer, Marconi operator, and a few other senior personnel lived aft in the main cabin which, compared with the forecastle and hold, was "palatial." They had their own cook and were attended by stewards. By contrast the men ate in their berths and had to fetch their food in tins from the galley. Their staples were hard bread, salt pork, and molasses-sweetened tea. Four days a week they received only tea and hard bread; on the other three they ate salt pork and "duff," a boiled pudding made from flour and water, with a bit of fat for flavour.[124] The men had small stoves called "bogies," on which they brewed "switchel," a crude tea made from loose tea leaves that never steeped, the water being at a constant boil.[125] The water, melted pinnacle ice, was regularly contaminated by rust and seal blood.[126] After the hunt got under way the men also used the bogies for roasting bits of seal meat, the only fresh meat they got on the voyage.

In such a working and living environment it is not surprising that injury, death, and disease were common. Contaminated water frequently gave rise to outbreaks of dysentry.[127] And the SS *Micmac* was forced to give up the hunt in 1874 and put into Greenspond when 40 of its crew came down with measles.[128] Not until 1909 was a regular system of medical inspection set up for returning sealers.[129] That year, when Newfoundland was in the middle of a smallpox epidemic, all the sealers were vaccinated and there were further vaccinations in 1910.[130] In spite of these precautions, in 1911 the SS *Newfoundland* returned to St. John's with 11 men suffering from the disease.[131] Scottish whaling ships of the prewar period all carried surgeons as a matter of course, but their example was not emulated by Newfoundland shipowners. Few sealing ships included doctors as part of their staffs — the men usually stitched wounds and set broken bones by themselves.[132] Those who died at the hunt and whose bodies could be recovered were put into rough coffins and preserved in salt.

Neglect of the men's safety had other manifestations. Some steamer captains did not have master's certificates and therefore brought along navigating officers. Usually, however, the captains ignored their advice. Charles Green, navigating officer of the SS *Newfoundland* in 1914, had been "explicitly told by the owners, Harvey and Company, that he was to have nothing to do with the running of the ship."[133] Lifesaving equipment was practically nonexistent. The 273 men of the SS *Bear* had only 25 punts at their service in the event that the ship went down.[134] Owners thought of the punts only in terms of their utility for reaching seals.[135]

The owners' disregard for the safety of the men reached its peak in 1914 when Alick Harvey of A.J. Harvey and Company, owner of the SS *Newfoundland*, removed the ship's wireless set because it was not paying for itself in the form of increased catches.[136] One member of the 1914 commission of enquiry into that year's sealing disasters concluded that the owners had viewed wireless solely as "a means of transmitting information as to where the main body of seals was to be found."[137] Thus, when the *Newfoundland*'s men became stranded on the ice in a blizzard, its captain, Wes Kean, had no way of knowing their whereabouts and mistakenly assumed they were safe on board the SS *Stephano*. As a result, 77 men perished after spending two days on the ice.

The only bright spot for the sealers in the prewar period came with the passage in 1893 of the so-called "Sunday law" that forbade killing seals on Sundays.[138] Prior to 1893 a minority of captains did not allow sealing on Sundays in observance of the Sabbath. But because not all captains were so devout, the "Sunday men" and their crews were at a disadvantage. Most sealers were opposed to Sunday work, their piety no doubt tinged by understandable self-interest.[139] The law, passed in order to place all the crews on the same competitive basis, was a concession to the Sunday men. Nevertheless, it benefited all sealers by giving them at least one day of rest a week at the ice.

In the aftermath of the *Newfoundland* and *Southern Cross* disasters in 1914, sealers' working conditions became a popular issue. Already political pressure from the FPU had led to the passage of the Seal Fishery Act on 11 March 1914, with many measures intended to take effect that season.[140] However, the new regulations were observed on only one ship, the *Nascopie*, and then only because FPU leader William Coaker was aboard for the voyage.[141] The act required significant changes in the sealers' diet. Hard bread, previously the staple on sealing vessels, was to be augmented by not less than one pound of soft bread for each man three times a week. Beef, pork, potatoes, and pudding were to be served for dinner three times a week, and at least once a week the beef was to be fresh or canned. Sunday soup was to contain onions, potatoes, and turnips. Stewed beans were to alternate with fish and brewis (hard bread soaked overnight) as breakfasts.

The act also addressed the key subject of living conditions on the ships, although these regulations pertained to steel ships only. Effective immediately, the interior of the hull that ran along the crews' sleeping quarters was to be sheathed with wood. Effective 10 March 1915 the crews' sleeping quarters were to be heated by steam pipes and furnished with iron-frame berths. The act also made some provision for improved medical care, but again this did not apply to wooden walls. One room on each ship was to be set aside for sick and disabled men, and each ship was to carry a doctor "wherever practicable."

Because the act was passed before the 1914 sealing season began, it did not address the abuses that contributed to the disasters that year. Those abuses became the target of subsequent legislation. That fall, new legislation made wireless sets mandatory on all ships.[142] A revamped Seal Fishery Act was passed on 4 May 1916 in obvious response to the disasters.[143] All seals were to be killed between daylight and dark, and no sealer was to remain on the ice more than one hour after dark. Captains would be fined if they sent men to hunt for seals at any time other than between daylight and dark, or if they sent them onto the ice "when the state of the weather is such as endangers life and limb." However, sealers could be put onto the ice after dark to help bring aboard pelts that had been panned during the day. Sections seven and eight of the act enjoined the owners to equip each vessel with fire rockets or flares, and instructed the captains to fire these and/or sound the steam whistle at regular intervals should any of their men remain on the ice "without lawful excuse after dark, or in fog, mist, or falling or drifting snow."

Section 13 reflected the consensus that the *Southern Cross* went down because of overloading. No ship was to return to port during a single season with more than 35 000 pelts; pelts over that number would be confiscated. The minister of Marine and Fisheries was empowered to survey all ships and to mark load lines on them. Any ship returning to port with its load line below water would be subject to a maximum fine of $2500.

As an aid to navigation, each ship was to carry a certified master or mate, although this requirement could be waived if the captain proved to customs officials that he could not find one. Similarly, section ten required the presence of a doctor on each ship carrying more than 150 sealers unless one could not be "obtained upon reasonable terms." In lieu of doctors, most vessels actually carried local pharmacists.

The 1916 act is particularly noteworthy because it was the first legislation to provide specifically for compensation to sealers in case of death or injury. Newfoundland had had a Workmen's Compensation Act since 1908,[144] but in its original form the act did not cover fishermen or loggers, the two largest occupational groups in the colony. Loggers were brought under the act in 1914, thanks to the efforts of the FPU, which elected eight members to the House of Assembly in the 1913 general election. The FPU was equally instrumental in extending the compensation principle to sealers, although no doubt the enormity of the sealing disasters made the job a little easier. Nevertheless, the gains were modest. Claims were limited to death or injury from exposure, and even this was qualified. The only men eligible for coverage were those who suffered from exposure as a result of failing to return to their ship within one hour after dark. The maximum claimable was $1000. There was no compensation for death or injuries due to explosions, drowning, infection, broken limbs, disease, or any of the host of other threats that the sealer faced on every trip. The compensation section of the act, while a step in the right direction, represented an extremely narrow application of the principle.

The last significant piece of pre-confederation sealing legislation was designed to give greater force to the regulatory provisions of the Seal Fishery Act of 1914. The Seal Fishery (Amendment) Act, passed on 13 July 1920,[145] forbade any steamer to clear for the seal hunt unless it carried a government-appointed inspector. The inspector was responsible for seeing that all food was up to standard and that the sleeping quarters of the crew were "kept at all times in a clean and proper condition." The inspectors appear to have had little effect. American journalist George Allan England, who went to the ice on the *Terra Nova* in 1922, concluded that the inspector "must have been a total paralytic, blind, deaf and otherwise disabled."[146] But even zealous inspectors were constrained by the narrow limits of their authority. They did not, for example, have any say in the storage and handling of gunpowder. Yet for all the comparatively limited scope of the changes, Captain Abram Kean lamented that the sealers had softened from too much "luxury," what with hot food to eat and real bunks to sleep in.[147] He longed for the days when sealers were "real men."

Disasters and near disasters were common as the wooden walls, already old, continued to age. Late in the afternoon of 14 March 1925, Farquhar Steamship Company's 43-year-old *Stella Maris* was struck by two ice pans and began to take on water, flooding the engine room.[148] The crew of 43 men made for the lifeboats but found them frozen to the davits. Just when it looked as though all hands would go down with the ship, the engineer coaxed a last burst of power out of the engines. Captain George Whiteley manoeuvred the ship alongside an ice floe and the men scrambled onto it. Only 20 minutes elapsed from the time the ship was struck until it sank. Fortunately, the steel ship *Prospero* received the distress call and rescued the crew after they had spent an anxious night on the ice. Of course, steel ships were not immune. On 9 April 1926 an explosion in the engine room of the *Seal* killed the oiler and chief engineer, and several men were badly burned. The survivors scampered off the ship and then watched as the fire ignited the ship's magazine, touching off a massive explosion that sent it to the bottom. Again, the crew spent the night on the ice before being picked up.

The last major disaster involving a wooden wall occurred in 1931. The *Viking*, then 50 years old, was carrying 147 men, including American film maker Varick Frissell and his crew, who planned to shoot extra footage for a movie set against the background of the seal hunt.[149] Frissell, like others before him, was alarmed by the sealers' indifference to gunpowder. On this trip, the ship carried more powder than usual because the film crew intended to stage a dramatic explosion for the cameras. Some of the extra powder was stored in the ship's bathroom, and men were known to knock their pipe bowls against powder cans or smoke cigarettes while they used the toilet. This carelessness is generally blamed for the explosions that rocked the *Viking* during the night of 15 March as it lay off the Horse Islands. Some sealers managed to jump to safety on the ice, but many were catapulted through the air by the force of the explosions. Twenty-eight men were killed, including Frissell, and countless others were maimed. No compensation was owed to the injured or to the dead men's families, since this type of accident was not covered by the compensation provisions of the Seal Fishery Act.

In spite of the risks inherent in using the deteriorating wooden walls, the sealing legislation did ameliorate some aspects of shipboard life for the men, especially their diet. Captain Bob Bartlett's recollection of food served on the sealing ships in 1929 demonstrates the changes, although some items from the past, such as duff, remained popular.

> *On ship we have what we call Solomon Goss's birthday. He has a birthday three times a week — Sunday, Tuesday, and Thursday. On these days for the noon meal we get duff. For duff, flour (a barrel or more to a batch) is stirred with water, currants, and molas-*

ses. With a blade as large as a canoe paddle, the cook mixes it into a paste; then he adds shortening — fat from boiled pork.

The dough is packed into canvas bags. These bags are two-men-duff size or three-men-duff size. The cook puts all the bags in a big boiler on the galley range and boils them for two to three hours. Another boiler alongside contains pork. When the duffs and pork are ready the cook calls the roll, and one man of each three messmates comes to the galley and gets the allotted share of pork and duff for himself and the other two men, waiting for him below in the quarters.

As the men come by, the cooks stab the duffs with miniature pitchforks, lift them out of the boiler and douse them in iced water. Then they deftly skin the bags off the duffs, which fall into the men's pans.

There are no tables or messboys to wait on the men.

On Sunday morning the crew gets "brose" — boiled bread and codfish with pork gravy spread over it. Butter, fresh beef, salt fish, potatoes, and turnips are whacked out to the men at different times, and they can prepare a meal for themselves any time they want to, except on Solomon Goss's birthdays, when the range is filled up with boilers.

Tea is made in big kettles, five gallons at a time. On duff days the men make their own tea with tea leaves the cook gives them or that they have bought with their crop money.

The afterguards have it better. The chief engineer, "Marconi" (radio man), second hand, barrelman, and doctor dine with the captain, while the officers dine in the mess room. The galley forward is not used by the afterguards at all.

When we get among the young seals we boil or fry the seal meat. A mess crowd will cook a bit of seal with onions and butter for themselves. It is good; sometimes, it seems to me, much better grub than the afterguard gets.[150]

The men's fare while they were on the ice also improved.[151] Some still carried hard bread in their canvas "nunch" or "nunny" bags, but it was now augmented with fruit (usually an orange), oatmeal sweetened with sugar, and perhaps a piece of seal meat or pork from the previous evening's meal. This change had been urged by the commissioners examining the 1914 disasters. They pointed out that hard bread had little nutritional value.[152] While acknowledging that some men did indeed take more nutritious oatmeal and raisins onto the ice, the commissioners had noted that the men themselves, and not the vessel owners, provided such supplements.

Processing

Like the ships themselves, the major processing facilities became concentrated in St. John's towards the end of the 19th century. And again, cost was the major factor. Until 1846 the seal vats in St. John's had been located on the north side of the harbour. That year they were all destroyed in a fire and thereafter the sealing merchants re-established their processing equipment on the opposite, less populous south side of the harbour in order to reduce the risk from fire.[153]

The initial stage of processing — separating the skin from the fat — changed little before 1914, continuing to be the domain of the seal skinners. Job Brothers alone employed 20 to 25 seal skinners.[154] The skinners, mostly St. John's butchers, performed their jobs on a piece-work basis and made $200 to $500 dollars each spring. A skinner's average output was 300 to 350 skins per day (nine to nine and a half hours). Once separated from the fat, the hides were stretched and salted ("pickled"), after which they were ready for export. At British and American tanneries the skins were placed in lime pits for several weeks to loosen the roots of the hairs, which were then removed to be used in making plaster and fertilizer.[155] By the 20th century the skin itself was split into two sections: the outer portion was tanned and made into finished leather products, while the inferior inner leather was used in manufacturing glue and fertilizer. The most important market by 1925, though slipping, was still the United Kingdom, accounting for 57 per cent of Newfoundland's total sealskin exports (149 222), the remainder going to the United States.

The process of rendering seal fat into oil was completely altered by the application of technology to what had been a natural process. The new technique had been developed by S.G. Archibald around mid-century.

> By means of steam-driven machinery, the fat is now rapidly cut up, by revolving knives, into minute pieces, then ground finer in a sort of gigantic sausage-machine; afterwards steamed in a tank, which rapidly extracts the oil; and finally before being barrelled, it is exposed for a time in glass-covered tanks to the action of the sun's rays. By this process the work of manufacturing, which formerly

occupied two months, is completed in a fortnight. Not only so, but by the steam process the disagreeable smell of the oil is removed, the quality improved, and the quantity increased.[156]

The various grades of seal oil, in descending order of value, were: (1) young harp, (2) young hood, (3) bedlamer, (4) adult harp, (5) adult male hood, and (6) adult female hood.[157] This was, however, subject to fluctuations, especially between young harps and young hoods. The waste from the rendering process was sold to local farmers, who mixed it with peat moss to make fertilizer.

For most of the period 1914–39, seal-processing facilities continued to be concentrated in St. John's. Competition emerged in Twillingate (Ashbourne and Company) and Port Union (Union Trading Company) during the interwar years in association with the landsman and auxiliary schooner operations, but the volume these plants processed was only a fraction of what was produced in the capital. In 1930, seal oil processed by Job Brothers and Bowring Brothers had a combined net value before export of $359 623.06, compared with $46 187.41 for Ashbourne and Union Trading.[158]

The first major change in processing after 1914 was the adoption of seal-skinning machines in 1926.[159] Experiments with mechanical skinners prior to that date had been unsuccessful. The new machines completely displaced the seal skinners, a result hastened by the decision of members of the Seal Skinners' Union not to work in any plant where the machines were also used.[160] The skinners were scornful of the quality of work done by the machines, but others, like Captain Bob Bartlett, felt the machines did a better and more efficient job.[161] From the owners' perspective, the obvious advantage was that, after the initial capital outlay, it was cheaper to operate the machines than to pay skinners' wages, the highest of any labourers in the industry. Manual labour was still required in the removal of bits of flesh, known as "tare" or "back weight," attached to the fatty side of the pelts. This work, called "fleshing," was performed for a pittance by young boys who thronged to the waterfront when the fleet returned.

In the 1920s each crew was responsible for unloading its own pelts after the ship docked at the southside premises. The pelts were weighed on large scales and the tare deducted to determine the paying weight. Once fleshed, the pelts were placed on a conveyor belt and carried to a mechanical skinning knife that separated the fat from the skin. Thereafter the fat was treated as it had been since the adoption of Archibald's process, being ground up and then reduced to a liquid state in pressurized vats called digesters. Final-

42 Sealing premises, south side of St. John's harbour, 1897.
National Archives of Canada, PA-38217.

43 Landing seal pelts at St. John's.
Provincial Archives of Newfoundland and Labrador, A18-31, Holloway Studio.

44 Skinning seal pelts, St. John's, *circa* 1900.
Provincial Archives of Newfoundland and Labrador, B1-103, Holloway Studio.

45 Skinning seal pelts, St. John's, *circa* 1930.
Provincial Archives of Newfoundland and Labrador, William C. Garland Collection, T.B. Hayward photo.

ly, the oil was placed in shallow tanks covered by hothouse glass frames and exposed to the sun for several weeks. The bleached oil was pumped into tankers for transportation to market. In 1925, 63 per cent of Newfoundland's 3208 tons of seal-oil exports went to the United States, the only other markets being the United Kingdom (30 per cent) and Canada (7 per cent). American demand for seal oil had taken off around the middle of the 19th century following the decline of American whaling; by the beginning of the 20th century the United States had overtaken Great Britain as the chief market.

There was at that time no comparable innovation in the treatment of sealskins, which were still cured in salt. However, a greater importance was attached to the skins. Sealskins were much sought by the fine leather trade in both the United States and Europe, and the price paid for them during the late 1920s, $2 to $3 each, was high.[162] A new factor was the interest European and American furriers showed in seal fur (technically hair), which they dyed to imitate more expensive animal furs.[163] This demand was evident by 1928. That year agents for several international firms were on hand in St. John's to greet the sealing ships as they arrived back from the hunt.[164] The agents paid up to $10 for the skins bearing the best fur at a time when $3 was considered an excellent price for a sealskin.

Pay

The key change in remuneration during the steamer era was that steamer crews shared only one-third of the proceeds of the voyage, as opposed to one-half on sailing vessels. The owners took two-thirds, from which they paid the captain at a fixed rate per seal.

At first the new sharing arrangement did not adversely affect the men's earnings, since in the 1860s and early 1870s the average catch per man rose because of the steamers' efficiency.[165] However, as the number of steamers grew and pressure on the resource increased, the average catch began to decline. In 1871 the 15 steamers in the fleet recorded an average catch of 124 seals per man. Ten years later there were 27 steamers and the catch per man had fallen to 56. This trend was reversed late in the 1890s, but only after legislation limited crews to 270 men, which reduced the total number of sealers. Also, by the 1890s the seal herds had become noticeably depleted, and this further contributed to a declining average catch. Overall, during the second half of the 19th century the men's earnings declined.

PLACING SEAL PELTS ON CONVEYOR OF SKINNING
MACHINE. BOWRING BROS LTD. ST JOHNS N.F. 1413

46 Seal pelts going onto the conveyor belt to the skinning machine,
Bowring Brothers' factory, *circa* 1930.
Provincial Archives of Newfoundland and Labrador, William C. Garland Collection, T.B. Hayward
photo.

47 Skinning machine, Bowring Brothers' factory, *circa* 1930.
Provincial Archives of Newfoundland and Labrador, William C. Garland Collection, T.B. Hayward photo.

48 Grinding seal fat, Job Brothers' factory, *circa* 1930.
Provincial Archives of Newfoundland and Labrador, William C. Garland Collection, T.B. Hayward photo.

49 Filling digesters with the seal fat, Bowring Brothers' factory, *circa* 1930.
Provincial Archives of Newfoundland and Labrador, William C. Garland Collection, T.B. Hayward photo.

50 Bleaching seal oil, Bowring Brothers' factory, *circa* 1930.
Provincial Archives of Newfoundland and Labrador, William C. Garland Collection, T.B. Hayward photo.

51 The sealers' strike, 1902. A.B. Morine, the lawyer representing the sealers, is standing in the carriage in the foreground.
Provincial Archives of Newfoundland and Labrador, A18-41.

The practice of cropping continued in the steamer era. However, the range of articles that constituted a crop was greatly reduced, being confined to a few items of use to the sealer only while he was at the ice. This was part of a trend in 19th-century Newfoundland by which merchants decreased their participation in the credit system and forced the state to assume a correspondingly larger role, manifested in the form of government relief payments.[166] The crop was limited to tobacco, perhaps a new pair of boots or a new knife, and occasionally a bottle of liquor.[167] The standard crop ranged from $9 to $12, but it was marked up an additional 33$\frac{1}{3}$ per cent if any seals were taken on the voyage, and rare was the ship that took no seals.[168] The amount of the crop was deducted from each man's share at the end of the voyage.

52 Captain Sam Blandford (1840–1909).
Provincial Archives of Newfoundland and Labrador,
C1-236.

Berth fees were abandoned in 1902 because of a strike that threatened that year's hunt. Several strikes against berth fees had occurred throughout the 19th century, including a particularly violent one in 1860, but they had never achieved lasting success. On 8 March 1902 three thousand sealers banded together in St. John's and issued four demands: (1) that the $3 berth fee be abolished; (2) that the price the owners paid for seal fat be increased from $2.40 per quintal (112 pounds) to $5, the international market value then being $6.50; (3) that the owners provide some food for the men between the time they arrived in St. John's and the time they boarded their ships; and (4) that the $33\frac{1}{3}$-per-cent markup on crops be discontinued.[169] For three days the owners refused to meet with the sealers' representative, lawyer A.B. Morine, and it appeared as though they might literally starve out the strikers, who were subsisting on handouts from local sympathizers. On the fourth day of the strike the owners made a counter-offer of a $1 berth fee

103

and $3.25 per quintal of seal fat. The men turned it down and sent Morine back to the owners. Later that day the owners offered no berth fee and $3.50 per quintal of seal fat, which the men accepted.

The largest "bill" (crew share) during the steamer era was achieved in 1866 by the men of the SS *Retriever*, owned by Ridley and Sons of Harbour Grace. They made two trips, taking 23 400 seals for a bill of $303 per man, a fortune at that time.[170] By 1911, with second trips a thing of the past, a sealer stood to make at most $80; in fact, between 1900 and 1914 the average annual share exceeded $50 only once (in 1910). From that the crop, say $12, had to be deducted. Also, each sealer had to pay his train fare or other travelling expenses to and from St. John's.[171] The abolition of berth fees in 1902 may have been offset by increased travelling expenses as the industry became concentrated in St. John's.

The captains, who received a certain amount per pelt, did very well. In 1874 a captain earned sixpence per pelt; by the late 1920s he took four per cent of his vessel's net earnings.[172] Abram Kean held the record for largest "bill" paid a sealing captain — $5433.94 for a 20-day voyage on the SS *Florizel* in 1916.[173] Perhaps more typical was the $1900 paid to Captain Henry Dawe of the SS *Adventure* in 1908.[174] Few people in Newfoundland other than merchants and professionals made more. It is indicative of their social and economic standing that at the Methodist church in Wesleyville, the sealing captains sat with the merchants in the seats closest to the pulpit.[175]

The relationship between the captains and their merchant masters did not begin and end with the sealing season. Merchants either employed the captains directly or used their influence to get them jobs. Captain Sam Blandford was superintendent of shipping for the Reid Newfoundland Company and managed Job Brothers' Labrador fishery stations.[176] Captain Thomas Fitzpatrick was appointed collector of customs at Placentia in 1910. The most successful sealing captain of them all, Abram Kean, who sailed mostly for Bowring Brothers, had the mail and passenger service contract for Labrador.

After 1914 the average share paid to sealers — based on the weight of the pelt — steadily improved. The average annual share for the period 1900–13 was $41.[177] During the First World War it rose to $76, a reflection of artificial wartime demand. In the interwar period, which included the Great Depression, the average share figures are somewhat misleading. For one thing, they hide extreme variations. In 1926, when the average share was $71, individual shares ranged from $12 for the crew of the *Viking* to $125 for the crew of the *Beothic*. The average share also does not account for extra expenses the sealers incurred, including crops and transportation costs.

Averaging shares obscured more than expenses. By the 1920s, if not earlier, some sealers were earning money selling seal flippers. The captains of a few ships made their men cut both fore flippers out of the pelt, but most allowed them to leave a single flipper, which became the sealers' property. In 1929, flippers sold in St. John's for 25 to 30 cents a "string" (dozen).[178] Based on that year's average catch per man (94 seals), the sealer earned approximately $2 from flipper sales. This hardly offset crop and transportation charges, but it did help.

The rising average share was not due to a new-found generosity on the part of the owners. It was largely a function of the smaller crews of the interwar years coupled with a rejuvenated seal population. Shares were also affected by the price that the owners set for seal fat. The so-called "price for fat" was actually a misnomer, since the pelts were weighed before the skins and fat were separated.[179] The owners automatically deducted the tare, which included the fur, from the weight of each pelt at rates ranging from $1\frac{1}{2}$ pounds for whitecoats to 7 pounds for old dog hoods, regardless of the actual amount.[180] In other words, the skins were lumped with the fat, and nothing at all was given for the fur. In the 1920s, as the market price for sealskins rose because of the increased demand for leather and fur, it became obvious that sealers were receiving nowhere near their customary one-third of the net proceeds of a voyage. For example, in 1927 the men's shares were based on a price of $4.50 per quintal of fat. This failed to account for the fact that the owners were getting an additional $2 to $3 for each hide and up to $10 for top-quality furs. The situation caused what one reporter euphemistically termed "restlessness" among the sealers.[181] In 1928 the owners staved off a strike by raising the price of fat to $5 per quintal, and the following year they jumped it to $5.50, again with strike action looming. Unfortunately for the men, the crews during these two years were the largest of the decade, preventing them from realizing exceptional shares. This suggests that crew size was the more influential factor.

Observers from outside Newfoundland were astonished by the pay in the sealing industry. George England had "never known a country where employers enjoyed such a sinecure as in Newfoundland. Labour, there, has hardly begun to dream that it has any rights. And the game of exploitation goes merrily on."[182]

Given the sealers' meagre earnings, why did men compete so eagerly to get berths on sealing ships? The primary motive was economic. The prospect of receiving cash was crucial, for all other branches of the Newfoundland fishery were based on credit. In 1933 the Newfoundland Royal Commission reported that "it could have been said with truth only a few years ago that there were families in Newfoundland who had never seen money in their lives."[183] Those who had were probably sealers' families. Although the sealer's pay was generally small, he at least had a chance of making what was, by his standards, a great deal of money; the analogy of the seal hunt as a lottery is appropriate.[184]

Men who went to the ice improved their credit rating with the local merchants and were, therefore, better able to provide for their families.[185] Moreover, during the six to eight weeks that he was at the ice, the sealer's food was provided by the vessel owner. This meant that in his home there was one less mouth to feed at a time when the family food supply was depleted after the winter. Captain Bob Bartlett summed it up best:

> *I have sailed for the North Pole and for the war and for a lot of other unusual destinations, but curiously enough there is no excitement to be compared with the sailing of a sealer out of St. John's, Newfoundland. The explorer or the soldier is going out to achieve something. But the sealer is going out to decide whether his wife and babies will have molasses cakes with their boiled codfish for the next eleven months.[186]*

Sealing was also important for cultural reasons. In the home communities of the sealers, participation in the seal hunt was viewed as the foremost rite of manhood.[187] The young man who returned from his first trip to the ice was accorded a new status, both within his own family and within the community at large. Conversely, those who did not want to go to the ice were viewed with scorn. Sport is a vital component of popular culture, and in Newfoundland sealing was the national sport. The enterprise was rich with competitions. The ships raced out of the harbours and raced to be the first to the ice. Once there, each crew tried to take the most seals. The work itself required strength, agility, and balance, all prerequisites for the athlete. The crews were like teams, the men endeavouring to serve on the same ship or under the same captain every year. There were also rivalries within each ship. The highlight after a day of shooting occurred when the gunners and dogs assembled in the main cabin for the count. Men gambled on who would be the "high liner," the winner gaining prestige and perhaps a drink with the captain. The biggest competition was between the crews to determine which ship would be the high liner of the fleet. The captain of the high liner received a silk pennant from the St. John's Chamber of Commerce, a tradition dating back to 1832.[188] Occasionally the owners offered a prize, sometimes a case of oranges, to the high-liner crew.[189] Such were the victors' trophies.

The Seal Hunt
Since 1939
An Overview

The Hunt
During World War II

During the Second World War the Newfoundland sealing industry almost disappeared. In 1939 the fleet consisted of seven steamers, but, just as in the First World War, most of the sealing ships were pressed into war service. Four steamers and one auxiliary schooner went to the ice in 1941, and the number of sealers dropped below a thousand for the first time since 1932. In 1942 there were only three steamers, all wooden walls, and one of these, the 71-year-old *Ranger*, went down in a storm, fortunately without loss of life. This was followed in 1943 by the loss of the intrepid *Terra Nova*, which was damaged by ice while ferrying supplies between Newfoundland and United States military bases in Greenland. The only Newfoundland vessel to participate in the 1943 hunt was the *J.H.* Blackmore of Port Union, one of the first vessels in the history of the Newfoundland sealing industry to be powered exclusively by a diesel motor (one small motor vessel had gone out as early as 1933). The 1944 hunt was left to the Bowring Brothers' wooden wall *Eagle*. The *Eagle* was back on war service in 1945 when the hunt was taken up by five motor vessels with crews totalling 114 men.[1]

The demise of the Newfoundland fleet during the war had several causes. The industry was probably deterred somewhat by the levy of five cents on every sealskin exported from Newfoundland between September 1942 and June 1945. Also, because the fleet was so small when war broke out, the diversion of even a few ships to war service was a major loss. Another important factor was profuse Canadian and American military spending during the war. So much money poured into the Newfoundland economy that the period was likened to "frontier days on the American continent."[2] Military base construction and maintenance created, at its height, approximately 19 000 jobs. Many of the people who filled these positions were former fishermen and sealers who willingly left their traditional calling for jobs that paid better and, moreover, paid in cash. The demise of sealing in this period contrasted with the buoyant cod fishery. Although salt cod exports did increase during the war, the major gains were in exports of frozen fish, which rose in value from $344 414 in 1939 to $5 864 038 in 1945. (Interestingly, one of the two leading firms producing frozen fish was Job Brothers.) The creation of employment on the bases and in the frozen-fish sector contributed to the wartime plight of the sealing industry.

In some ways where the sealing industry was concerned, a lot of good came out of the war. The appearance of motor vessels sparked a transformation of the industry. Their numbers rose to 11 in 1946, 15 in 1947, and a historic high of 21 in 1948. The rise of the motor vessel broke the St. John's monopoly of the industry. Motor vessels hailed from several outport centres, including Catalina, Port Union, Twillingate, Little Bay Islands, and Channel–Port aux Basques. In 1948 nine firms or individual owners were represented in the hunt.[3] Unlike the steamers, most motor vessels (10 out of 12 in 1951) were built in Newfoundland, many of them at the government's encouragement as part of its effort to modernize the fishing industry.[4] Although intended for the cod fishery, motor vessels were adaptable to sealing. The advent of the motor vessel, therefore, held out the promise of a return to economic prosperity for the outports, based on the allied enterprises of sealing, cod fishing, and shipbuilding.

One other factor was cause for optimism. The seal herds had been virtually unmolested during the war and there was every reason to believe that they had increased. In addition, because the number of sealers was smaller after the war, there was less pressure on the herds. The 4 steamers and 12 motor vessels at the hunt in 1948 carried only 1035 men, yet this was the highest number since 1940. Using marine biologist J.S. Colman's measure of 4000 sealers as the danger limit for overharvesting, there was plenty of room for expansion. Since the average motor vessel of the immediate postwar period carried 25 sealers, there would have to be 160 vessels for stocks to be threatened. In his 1949 review of the Newfoundland sealing industry, Colman concluded that "there is not likely to be any danger to the Newfoundland seal-herds during the next 10 years."[5]

Postwar Decline

The glowing future that Colman predicted for the industry never materialized. By 1950 the sealing fleet was reduced to four motor vessels. There was a short-lived revival in 1951 and 1952 (12 and 11 motor vessels respectively), but by 1954 the fleet was reduced to three motor vessels with total crews of 207 men. For the remainder of the decade, the fleet ranged from a high of five motor vessels in 1957 to a low of one in 1959. The problem lay not with the price of seal oil, which matched or exceeded price levels of the Second World War, but with the price paid for sealskins, which failed to increase in the 20 years after 1938.[6] Since skins now contributed as much as 60 per cent of the value of the catch, this had a profound effect on profits.

The 1950s, like the 1920s, were key years in the history of the industry. Newfoundland's confederation with Canada in 1949 brought with it a host of social benefits that undermined the traditional work pattern in the outports.[7] It was no longer so important for fishermen to risk their lives at the seal hunt every spring for uncertain reward. The average family allowance in 1949 was $16.38 a month, or nearly $200 a year.[8] The incentive for participating in the hunt was further reduced in 1957 when fishermen became eligible to receive unemployment insurance benefits. In fiscal year 1957–58, unemployment insurance payments to fishermen amounted to $1 759 000.[9] Two years later the figure had almost doubled. That these and other social benefits were even considered as substitutes for earnings from sealing speaks volumes about the pay in the industry.

The 1950s paralleled the 1920s in that the sealing industry suffered from government and entrepreneurial neglect. Premier Joseph Smallwood's government devoted most of its energies to an ill-fated effort to develop a manufacturing sector. This program, which cost the people of Newfoundland $30 million, was abandoned in 1957.[10] Provincial aid to the fishing industry took the form of $13 million in loans to the private sector for the erection of plants for processing fresh frozen fish.[11] This meant further neglect of the salt-fish sector on which the fishing industry in Newfoundland had historically been based. The value of frozen-fish production surpassed that of salt fish in 1955 for the first time ever.[12] The south coast bank fleet disappeared in this decade and the Labrador floater fishery — which also produced salt fish almost followed suit. Given the fact that the government overlooked the salt-fish sector, it is hardly surprising that the far less important sealing industry was also ignored.

The Newfoundland government's approach to fisheries management was matched after 1949 by the federal government. During the 1940s, federal fisheries officials had decided that the Maritime provinces fishery would concentrate on producing fresh and frozen fish for American and central Canadian markets.[13] When this vision of fisheries development merged with that of the Smallwood government in the fifties, the consequences were disastrous. By emphasizing continental markets, the federal government ignored Newfoundland's salt-fish sector, and thus stood idly by as the traditional salt-cod bank fishery lapsed into oblivion. Federal policy took no account of Newfoundland's traditional trading partners, the salt-fish-consuming countries of the Mediterranean, the Caribbean, and South America. Newfoundland's overseas trade was replaced by an unhealthy dependence on virtually a single market for frozen fish, the United States. When that market became saturated in the mid-fifties, the result was a full-scale depression in the Newfoundland fishing industry.[14] While the sealing industry did not go the way of the bank fishery, it came dangerously close.

The post-confederation Newfoundland sealing industry also continued to be characterized by a lack of business initiative. Job Brothers withdrew from sealing altogether after the 1952 season to focus on frozen fish. In 1958 Bowring Brothers announced that it was selling its two remaining motor vessels and getting out of sealing. Derek Bowring, one of the firm's directors, explained: "It's simply a question of economics; the cost of maintaining the seal hunt is way out of proportion to the returns."[15] Bowring complained that the firm had made "only" $5000 profit in 1956, when both vessels secured "bumper crops." Since the 1957 season was a failure, observers assumed that Bowring Brothers had suffered substantial losses. As it turned out, the announcement of the firm's withdrawal was premature. Reports of large patches of seals prompted a last-minute decision to send the *Terra Nova* to the ice for the 1958 hunt, and Bowrings remained in the industry for a few more years. But the withdrawal announcement was an indication of the firm's declining commitment to the industry.

As the Newfoundland firms scaled down or ended their participation in the seal hunt, they left a void that was filled by Nova Scotian and Norwegian entrepreneurs. Actually, it was difficult to distinguish between the two. The leading Nova Scotian firms, Karlsen Shipping Company Limited and Christensen Canadian Enterprises, had simply transferred their operations from

Norway to Canada while retaining their Norwegian ownership. In 1951 five Nova Scotian motor vessels took part in the hunt (mainly in the Gulf of St. Lawrence), compared with 12 from Newfoundland.[16] By 1954, however, there were more Nova Scotian vessels than Newfoundland ones, a lead that was not relinquished until the late 1970s.[17] Any consolation in the Nova Scotian takeover lay in the fact that the Nova Scotian firms looked to Newfoundland for their crews. Nova Scotian ships bound for the Front usually cleared from St. John's, where they took on their crews.[18] Newfoundlanders who wanted to serve on Nova Scotian vessels in the gulf travelled all the way to Halifax to do so, lured by higher pay than they would receive in the Newfoundland fleet. At the end of the season, Newfoundland crewmen were dropped off either at St. John's or Channel–Port aux Basques, the western railway terminus. The vessels then made for Nova Scotia with their cargoes.

Norwegian-based sealing companies overcame much greater distances than those faced by their countrymen in Nova Scotia. Norwegian vessels returned to the Front in 1946 not only because of the promise of their earlier experience, but also with the added incentive that the Soviet Union now denied them access to the depleted White Sea harp stock.[19] Throughout the 1950s, Norwegian participation ranged from 10 to 13 vessels annually.[20] By the end of the decade the Norwegian fleet outnumbered the entire Canadian contingent (Nova Scotia and Newfoundland). The Norwegian record, more so than Nova Scotia's, presents a striking contrast to the industry's bleak performance in Newfoundland. Even with all the advantages bestowed by its proximity to the seals' migratory routes, Newfoundland failed to compete with its rivals.

Newfoundland's decline continued through the 1960s despite rising prices for sealskins following upon improved methods of treating seal pelts to keep the fur fast to the skin. In 1961 three owners were represented at the ice: Bowring Brothers (*Algerine*); Captain J.H. Blackmore of Port Union (*Saint Addresse*); and Earle Freighting Services of Carbonear (*Finback* and *Arctic Eagle*).[21] Blackmore withdrew in 1962, leaving Bowring Brothers and Earle Freighting, which each sent one vessel to the ice that year. They were joined in 1963 by Newfoundland Engineering and Construction Company Limited of St. John's, owners of the *Sir John Crosbie*. Newfoundland Engineering and Construction was a subsidiary of Crosbie and Company, which occasionally had outfitted sealing ships earlier in the century. Together these three firms sent four vessels to the ice in 1963 and 1964, and three in 1965. Earle Freighting pulled out after the 1966 season and Bowring Brothers after the 1967 hunt. That year there were only two Newfoundland vessels, compared with nine from Halifax, six of them Karlsen ships.[22] The *Chesley A. Crosbie*, owned by Chimo Shipping Limited, another Crosbie subsidiary, was the only Newfoundland vessel from 1968 until 1970, when it was joined by the *Lady Johnson*, owned by Captain Clayton M. Johnson of Catalina. These two again represented Newfoundland in 1971, but in 1972, for the first time in over a century and a half, no Newfoundland ships sailed for the hunt.

Surprisingly, the plight of the industry stirred little apparent public interest within Newfoundland. The standard explanation was that Newfoundland sealing vessels were not modern enough.[23] Earle Freighting's *Terra Nova*, which went to the ice in 1963 and 1964, was the old Bell Island ferry *Maneco*, and the *Kyle* a pre-confederation coastal steamer. During the 1965 season the *Kyle* burned the 700 tons of coal it had taken on at St. John's and had to be refuelled on the homeward voyage.[24] The only truly modern vessels of the lot were the *Sir John Crosbie* and *Chesley A. Crosbie*, ice-strengthened and built with sealing in mind. Beyond this obvious weakness, however, lay only bewilderment and frustration, as a St. John's newspaper editorial illustrates:

> *Despite all the reasons that have been given out during the past ten or a dozen years to explain why the "Newfoundland" seal hunt has become a misnomer, many people are still puzzled by the fact that not only is the foreign sealing fleet increasing, but so is the Canadian sealing fleet, at its new base in Halifax.*[25]

Pathetic proof of this frustration came in 1967 when Bowring Brothers entered into a joint venture with a Norwegian firm and hired a Norwegian as captain of the *Algerine*, "to see if the Norwegians know something about the hunt that we don't."

For outport firms like Earle Freighting, replacing older vessels with expensive modern ones probably was a problem. But Bowring Brothers clearly possessed the capability to modernize. Unfortunately, it lacked the will. By the 1960s, shipping was becoming less and less important to the firm's Liverpool parent, C.T. Bowring, which was preoccupied with the more lucrative insurance business (C.T. Bowring, purchased by American insurance giant Marsh and McLellan in 1980, finally got out of shipping altogether in 1983). Quite simply, the firm no longer needed the relatively small profits it derived from sealing, and few other Newfoundland entrepreneurs were willing to fill the void. The industry's dismal performance in the 1960s was the direct result of the neglect it had suffered in the previous decade and, arguably, since the First World War. Inadequate sealing ships were the symptom, not the disease.

Conservation

Canada and Norway more than compensated for Newfoundland's reduced presence on the ice floes. By 1961, 28 Canadian (including four Newfoundland) and Norwegian sealing vessels were operating in the Gulf of St. Lawrence and at the Front.[26] According to J.S. Colman's earlier calculations, this number should not have threatened the seal population. However, Colman had been thinking in terms of the small Newfoundland motor vessels of the immediate postwar period and overlooked Canadian and Norwegian participation. As early as 1949 the combined catch of the Canadian and Norwegian fleets was double prewar levels.[27] Colman also failed to consider the improved hunting efficiency of the motor vessels. While not as powerful as some steamers, they enjoyed certain advantages: superior navigating equipment, including radar; assistance of government icebreakers; helicopters to put men onto the ice and to pick up pelts; an improved aerial seal-spotting service; and the ability to remain at sea for longer periods, thanks to shipboard refrigeration facilities for pelt storage. The average annual catch from 1949 to 1961, excluding the landsman take, was 255 000 seals. By contrast, the average annual catch of the Newfoundland steamer fleet from 1929 to 1939 was 156 000. Nor is 255 000 an absolute figure since official statistics do not account for pelts lost from panning, or for seals shot but never recovered.

Everything pointed to the undeniable fact that the harp and hood seals of the northwest Atlantic were under tremendous strain, with most of the pressure on the harps. (From 1895 to 1946, hood seals accounted for only four per cent of all seals taken, and by 1955 they made up less than one per cent.[28]) The rapid increase of the east coast sealing fleet after the war prompted the Canadian government to convene a meeting at St. John's in December 1949 to discuss possible conservation measures.[29] In attendance were sealing captains, owners, representatives of the Fisheries Research Board of Canada, and other representatives of the Canadian Department of Fisheries. They agreed that further information on the seal population was required before specific responses could be planned.

The population study that resulted from the 1949 meeting concentrated on the harp seal stock, which was felt to be in the greater danger. Conducted under the direction of Dr. H.D. Fisher and Dr. David Sergeant of the Fisheries Research Board of Canada, the study relied on aerial photographic surveys. The teams carried out aerial photographic surveys of the Front and gulf herds in March 1950 and March 1951 and based population estimates on those surveys.

After locating a whelping patch, its area was determined by flying the length and breadth at several points at known speed. Photography was carried out at from 200 feet to 3,000 feet altitude depending on light conditions, nervousness of the seals, and the regularity of their distribution on the ice. A strip was flown diagonally across the patch, with the automatic camera photographing every few seconds. The area of each exposure is known, and the total area of the exposures was determined. The adults and pups were counted (the latter from telephoto shots at altitudes of over 600 feet) and the density per square mile was calculated. From this and the total area of the patch (usually about 50 square miles east of Newfoundland) the total population of the pups could be estimated directly.[30]

The method left much to be desired. Adult males were generally absent from whelping ice; adult females feeding under the ice could not be counted, nor could many whitecoats, whose colour made them difficult to see against ice and snow. At the very least, however, the survey did establish relative population size and provide a bench mark for future work.

The researchers estimated the northwest Atlantic harp seal population to be 3 300 000. Of that number, approximately 645 000 were pups (215 000 in the gulf and 430 000 at the Front). Fisher felt that the existing kill level of pups could possibly be maintained as long as the breeding or adult stock was adequately protected. But he saw a major problem. Since the catch of breeding stock was double the prewar level, fewer pups would be born in the future. The increased pressure on breeding seals came largely from the Norwegians, who followed the herds north along their migration route, hunting them well into May.[31] Fisher suggested that a closing date of 30 April would offer a measure of protection to breeding females. Canada and Norway did reach a gentlemen's agreement in 1952, but it dealt only with opening dates, set at 5 March for the gulf and 10 March for the Front.

Prior to the 1955 season, industry and Department of Fisheries officials met in Halifax to discuss Fisher and Sergeant's findings. Industry representatives were reluctant to accept any changes, so the only agreement was that further studies were necessary. Specifically, another aerial survey was proposed to substantiate the 1950–51 census, and no closing date would be imposed "until such time as scientific data can be produced which will show beyond any reasonable doubt that there is need for such a restriction."[32]

The next survey was not carried out until the 1959 and 1960 seasons. Led by Dr. David Sergeant, the Fisheries Research Board team calculated that the harp pup population had declined from 645 000 in 1950–51 to 365 000 (150 000 in the gulf and 215 000 at the Front).[33] The total harp population, excluding pups, was estimated to be 1.25 million. Sergeant stressed that the dangerous trends Fisher observed in 1950–51 were still present. Pups, immatures, and adults were all being overharvested. For example, the catch of adults and immatures had tripled since the war, the adults bearing the brunt. Alarmingly, the proportion of immatures in the adult/immature catch had decreased from 75 per cent in the 1930s to 20 per cent in the 1950s, reflecting an "excessive" catch of pups. Sergeant concluded that "either the catch of immatures and adults should be greatly reduced in order to increase the breeding stock, or the catch of young should be considerably reduced in order to maintain recruitment." The industry could not have it both ways if it wished to survive.

The population surveys demonstrated that regulation of the seal hunt was imperative. But before regulation could be made effective, a major obstacle had to be overcome. The Front, where the Norwegians concentrated their hunting effort, lay in international waters. In 1961 the Canadian government proposed to the International Commission for the Northwest Atlantic Fisheries (ICNAF) that it assume responsibility for the conservation of harp and hood seals outside Canada's three-mile territorial sea. ICNAF subsequently drafted a harp seal and hood seal protocol for the International Convention for the Northwest Atlantic Fisheries. To become effective, the protocol had to be ratified by the 13 countries that were signatories to the convention, which proved to be a long time coming. However, in 1961 Canada and Norway did at least agree to end their hunts on 5 May. Sergeant viewed the agreement as "a compromise between the advice of scientists and the need of industry to make paying voyages."[34]

By 1961, groups and individuals outside government had begun to express concern about the harp seal population. Newfoundland journalist Harold Horwood decried the wastefulness associated with the hunt.[35] He recommended that sealing should be forbidden in May to protect the breeding females; that panning should be prohibited; that all killing of adult seals should cease; and that Canada and Norway should limit the number of sealing vessels and impose catch quotas. Among conservation groups, the Canadian Audubon Society took the leading role, drawing attention to the hunt by means of articles and editorials in its publication *Canadian Audubon*. Indeed, Horwood's article, which appeared in *Canadian Audubon* in 1960, was the first to draw wide public attention to the harp seals' plight. It should be stressed, however, that both Horwood and the Canadian

Audubon Society owed much of their awareness to the work of federal
Fisheries Research Board scientists.

The Rise
of the Protest Movement

Interest in harp seal conservation was matched by concern over killing
methods. The subject was first broached in 1949 by Dr. Harry Lillie, a Scottish surgeon and conservationist who had long been critical of whalers' techniques. Lillie witnessed the 1949 seal hunt when he accompanied the Newfoundland fleet as medical officer on the MV *Codroy*. Whitecoats

> *were generally killed quickly by a blow on the head, but oc-
> casionally I saw men in a hurry just daze them with a kick and cut
> the little bodies out of the pelts while they lay on their backs still
> crying.*[36]

Lillie was equally appalled by the wastage that for so long had been part of
the adult seal hunt:

> *Some seals died at once, others, shot through the neck or lungs,
> writhed on the pans until they flopped over the edge into the water
> to die out of sight. I saw as many as five seals from one single large
> ice pan disappear, leaving five trails of blood.*[37]

In St. John's following the hunt, Lillie met sealing captains and representatives of the owners. Many of them seemed sympathetic, but if they really were, they were a long time proving it. Lillie subsequently travelled to
Ottawa for discussions with officials of the Department of Fisheries. There
he was encouraged to learn that the department was already contemplating
its 1950 population survey.

Perhaps disillusioned by government inaction on the cruelty issue, Lillie
returned to the ice floes in 1955.[38] He filmed the hunt and afterwards distributed copies of his film to humane societies throughout North America.
Meanwhile, on the other side of the Atlantic, the British public became
aware of the issue upon publication in 1955 of Lillie's book *The Path
Through Penguin City*, which contained his account of the 1949 seal hunt.
Although neither the film nor the book provoked widespread public reaction, the issue had been brought into the open for the first time.

The cruelty issue exploded onto the international scene in 1964 after the
Canadian Broadcasting Corporation's French-language television network

aired a film on the seal hunt, *Les Phoques de la Banquise*, as part of its series on hunting and fishing in Quebec. The film, produced by Artek Films Limited of Montreal, contained footage on the Magdalen Islands landsman hunt, including one horrifying scene in which a seal was skinned alive and its carcass left flailing on the ice. Understandably, when the film aired in Quebec it caused an uproar. It inspired Peter Lust, editor of the Montreal German-language weekly *Montrealer Nachrichten*, to write an article entitled "Murder Island" (referring to the Magdalens), which was picked up by the German press. Eventually, it appeared in newspapers and magazines across Europe. Suddenly the Canadian government was faced with an international protest.

The outcry forced the government to take action. After consulting representatives of the sealing industry, humane societies, and the Canadian Audubon Society, it introduced the Seal Protection Regulations on 29 October 1964.[39] The regulations (1) established a quota of 50 000 harp seal pups in the Gulf of St. Lawrence; (2) made it a crime to skin a live seal; (3) established guidelines for clubs, which could not be less than 24 inches long and 1.75 pounds in weight; (4) prohibited the killing of adult seals in breeding patches, where a number of adult females were taken incidental to the whitecoat catch; (5) outlawed panning overnight except under special circumstances, such as storms; (6) established licences for vessels over 30 feet in length and permits for those under 30 feet; (7) required the licensing of all sealers; (8) restricted the use of aircraft at the Front to spotting only; (9) established seasons of 12 March–30 April at the Front and 7 March–25 April in the gulf, the latter subject to change at the discretion of the minister of Fisheries. The opening and closing dates did not apply to landsmen. Ironically, the incident in the Artek film, largely responsible for bringing about these improvements, had been staged. A Magdalen Islands fisherman later gave a sworn statement that he had been paid to skin a live seal, and moreover, that he had done so on 3 March 1964, before the sealing season had officially opened.[40] In spite of this, the regulations were timely and progressive, representing what *Canadian Audubon* termed "major concessions on the part of the sealing industry."[41] Further progress was achieved when Norway agreed informally to keep its sealing vessels out of the gulf in 1965.

It was one thing to pass regulations, quite another to convince the public, national and international, that they worked. To that end the Canadian government arranged for a team of observers to witness the 1966 gulf hunt. The team consisted of Mr. Tom Hughes, general manager of the Ontario Humane Society; Dr. Norman Scollard, curator of Toronto's Riverdale Zoo;

Dr. Douglas H. Pimlott, professor of zoology at the University of Toronto and a director of the Canadian Audubon Society; Mr. John Walsh of the International Society for the Protection of Animals; Mr. Jacques Vallée of the Canadian Society for the Prevention of Cruelty to Animals (SPCA); Dr. Forbes MacLeod of the Saint John, New Brunswick, SPCA; veterinarian Dr. Elizabeth Simpson of the New Brunswick SPCA; and Mr. Brian Davies, executive secretary of the New Brunswick SPCA. With the exceptions of Simpson and Davies, the group issued a joint statement that generally supported the hunt, but questioned killing methods.[42] The majority of carcasses examined had crushed skulls, indicating that those seals were dead before skinning. However, some had not sustained any skull damage and team members "were unable to satisfactorily determine whether these animals had been rendered unconscious by any other means prior to skinning." They therefore recommended that the Department of Fisheries seek alternatives to clubbing. The team acknowledged the role of sealing in the culture and economy of the region, and accepted its continuance "provided that adequate regulations can be drafted and enforced to ensure that all seals are killed or rendered insensitive to pain by a humane method before skinning."

The subject of humane killing increasingly dominated public debate. The humane movement itself became split on the issue, one side favouring total abolition of the seal hunt, the other seeking to make it more humane. The abolitionists were led by Brian Davies, a Welshman who emigrated to Canada in 1955, serving briefly in the Canadian army before joining the New Brunswick SPCA in 1961. Davies first became interested in the seal hunt on 20 May 1964 when he attended a meeting of government, sealing industry, and humane society representatives in Moncton. During the course of the meeting he became convinced that "an overcapitalized sealing industry was intent on killing the last seal pup in order to get a return on its equipment."[43] On 13 March 1965 Davies witnessed the gulf seal hunt for the first time. The sight of men clubbing seals sickened him. "What right had man to treat these beautiful and intelligent animals in such a manner?"[44] What happened when he returned to shore on Prince Edward Island was probably more important than what he saw on the ice. Two whitecoats had been brought ashore and abandoned by hunters. Christening them Jack and Jill, Davies took the seals back to his Fredericton home. To purchase food to keep the pups alive, the New Brunswick SPCA established the Save the Seals Fund. Jack and Jill were subsequently transferred to the aquarium at the Vancouver Zoo, where they both died within the year. Meanwhile, the Save the Seals Fund was maintained to provide the financial basis for the New Brunswick SPCA's anti-sealing campaign. Davies withdrew from teachers college to oversee the campaign.

Davies and Dr. Elizabeth Simpson had exempted themselves from the joint statement issued by the observers of the 1966 hunt. Simpson claimed that although 95 per cent of seal skulls were crushed when federal fisheries officers were present, only 50 per cent were crushed when they were not. To Davies the "most distressing part of the hunt" was watching mother seals lose their pups, after which they kept "a silent vigil before the frozen bodies of their young."[45] After that year's hunt, Davies flew to England to meet Dr. H.C. Bernhard Grzimek, a prominent West German conservationist and critic of the hunt. This was part of Davies' strategy to arouse public opinion against the hunt in both Canada and Europe. Europe was especially important because European countries were the main buyers of seal products. Davies was more immediately successful in Canada, where an article he had written for *Weekend Magazine* in the spring of 1966 became his first major publicity coup. In it Davies called for the creation of a seal sanctuary in the Gulf of St. Lawrence to attract tourists interested in watching the seals.

Although Davies visited Europe again after the 1967 hunt, his publicity breakthrough there did not occur until 1968. That year he headed a delegation of 18 observers at the gulf hunt, including a photographer and a reporter from the London *Daily Mirror*, the most widely circulated daily newspaper in Great Britain. As anticipated, the *Daily Mirror* ran a sensationalist story on the hunt, accompanied by a photo of a sealer about to kill a whitecoat. The New Brunswick SPCA was soon inundated with letters and financial support from Great Britain. The inclusion of a reporter and a photographer from *Paris-Match* in the 1969 entourage stirred up similar indignation in France.

It was during 1968 that the first element of disingenuousness crept into the Davies campaign. Dr. Grzimek had agreed to co-operate with the West German fur trade in sending a group of veterinary pathologists to the gulf "to determine the effectiveness, or otherwise, of the present killing methods." These were Dr. Charles F. Helmboldt (United States of America), Dr. Bruno Schiefer (West Germany), Dr. H.C. Loliger (West Germany), and Dr. Lars Karstad (Canada), who was a special consultant to the Save the Seals Fund of the New Brunswick SPCA. Since Grzimek was Davies' chief financial backer at this point, Davies assumed responsibility for the pathologists' safety while they were on the ice. Before the hunt began, Davies did a remarkable thing. As he wrote in *Savage Luxury*,

Although I believed the seal hunt was primarily an ethical issue, I was quite happy to have pathologists provide me with scientific facts. This was assuming, of course, that they would limit themselves to fact. It had been my experience in the past that some veterinarians ... insisted on larding their reports with personal opinion.[46]

He therefore sent a memorandum to the pathologists, requesting that they confine their reports "to a presentation of autopsy findings. With regard to public statements made during and after the hunt, I would ask that the veterinary pathologists not state that the harp seal hunt ... is, or is not cruel." This amounted to telling them that they were not to draw conclusions from their data, and also implied that a layman such as Davies was better able than trained medical men to assess what constituted cruelty. Davies had cause to be uneasy. Although the Schiefer and Helmboldt reports are not available, the Loliger and Karstad findings are.[47] Of the 361 carcasses they examined, 349, or 96.7 per cent, were deemed to have been unconscious prior to skinning. Davies stated that, in spite of the obvious improvements since 1966, the report showed that "some of the animals were killed under highly questionable circumstances."

The hunt had indeed become more humane, thanks to the efforts of the Canadian government and humane societies not in the abolitionist camp, notably the Saint John, New Brunswick, SPCA and the Ontario Humane Society. Tom Hughes of the Ontario Humane Society was an expert in the field, having been active in the campaign for more humane killing methods in Canadian slaughterhouses in the early 1960s.[48] After the 1966 seal hunt the Ontario Humane Society had conducted its own investigation into seal-killing methods, even experimenting with electric shock. They discovered that clubbing was the most humane technique. To improve the effectiveness of clubbing, the Canadian government introduced stricter club specifications for the 1967 hunt. It also forbade the use of the gaff because of reports that sealers sometimes stuck gaff hooks into live seals.[49] Finally, the government increased surveillance of the hunt by fisheries inspectors, who could revoke the licence of any sealer guilty of inhumane killing practices.[50]

Another key event in regulating the hunt took place in 1966 when Italy, the last outstanding member country, signed the ICNAF Harp Seal and Hood Seal Protocol. This cleared the way for formal international regulation of sealing operations at the Front. The first step in this direction was changing the closing date for the 1968 hunt from 30 April to 25 April, further reducing pressure on the breeding seals. Also in 1968, the opening date at the

Front was shifted from 12 March to 22 March to ease pressure on whitecoats. These measures were influenced by the most recent seal census, conducted in 1964 under the leadership of Dr. David Sergeant. Sergeant found that annual production of pups at the Front had declined from 430 000 in 1950–51 to 200 000.[51] Even more alarming was the news that in 1964, Canadian and Norwegian sealers at the Front took 85 per cent of the entire pup production compared with an estimated 53 per cent in the gulf. At that rate the Front herd faced extinction. ICNAF action was designed to prevent this from happening.

Altering the opening and closing dates at the Front was a compromise. In 1967 the Canadian Audubon Society had argued for either a quota of 80 000 pups and 20 000 adults, or a 50-per-cent reduction in the Canadian and Norwegian fleets plus a 25 April closing date.[52] But ICNAF proved reluctant to set quotas. To its credit, the Canadian government acted decisively in the Gulf of St. Lawrence, which was actually less in need of protection. It banned hunting whitecoats there in 1970 and in 1972 closed the gulf to vessels more than 65 feet long. This had the effect of reserving the gulf for landsmen; it also meant that the vessel-based hunt now focussed exclusively on the Front. In another effort to placate critics, the government appointed the Committee on Seals and Sealing (COSS) in 1971 to advise the minister of Fisheries on economic, sociological, ecological, and humanitarian aspects of sealing. It was a COSS recommendation that led to the banning of large vessels in the gulf in 1972.

The Front herd got a reprieve in 1970 when Canada and Norway agreed to a 22 March opening date, permitting a major escape of young seals.[53] Canada and Norway also agreed not to take any whitecoats that year. In 1971 ICNAF finally introduced the first quota under the Harp Seal and Hood Seal Protocol. The total allowable catch (TAC) of 245 000 harp seals was broken down into 200 000 for vessels (100 000 each to Canada and Norway) and 45 000 for landsmen. Additional research indicated that this level was far too high, so in 1972 the TAC was reduced to 150 000, of which 120 000 was allotted to vessels (60 000 each to Canada and Norway) and 30 000 to landsmen. ICNAF scientists believed that at this level, stocks could rebuild. The 1972 TAC was retained through the 1975 season.

The flurry of new sealing regulations and the initiation of quotas for the Front failed to blunt the protest movement. Brian Davies had left the New Brunswick SPCA in 1969 because some members felt the association was neglecting animal welfare in New Brunswick at the expense of the sealing issue.[54] He was allowed to take with him the assets of the Save the Seals Fund, which he used to establish the International Fund for Animal Welfare

(IFAW). Within four years IFAW was generating revenues in excess of $500 000.[55] In 1971 both IFAW and the European Committee for the Protection of Seals offered to pay sealers not to hunt.[56] Three years later IFAW hired the New York advertising firm McCann-Erickson to co-ordinate its "Stop the Seal Hunt" campaign.[57] (The firm's corporate clients included the Coca-Cola Company.) Southern Ontario was saturated with newspaper, billboard, and television ads. Although the campaign cost IFAW over $100 000, this was more than covered by the fund's increased revenues, which leapt from $513 334 in 1973 to $805 141 in 1974.[58] IFAW continued to push for the creation of a marine park in the Gulf of St. Lawrence. To prove the scheme's viability, Davies transported tourists and students to and from the ice in 1970 and 1971. Nor did Davies lose sight of the European audience. The IFAW media entourage at the Front in 1973 focussed primarily on the Norwegian fleet, which consisted of 11 ships to Canada's seven. Davies explained that the shift in emphasis had come about because "the Canadian industry really doesn't exist anymore."[59]

Thanks to IFAW, the issue of humane killing had come to dominate the seal hunt debate, but in 1976 the dormant conservation issue was revived as the result of a study that challenged accepted wisdom on the northwest Atlantic harp seal population. In 1974 Dr. David Lavigne and Dr. Nils Oritsland, zoologists at the University of Guelph, discovered that ultraviolet photography could provide a more accurate count of harp pups.[60] The harp pup's white coat absorbed much of the ultraviolet light in sunlight; since snow reflected most of the ultraviolet light, the pups appeared in ultraviolet photographs as black spots on a white background, making them easier to count. Lavigne and Oritsland surveyed the Front and gulf herds in March 1975 and their findings were sobering. Their maximum estimate of pup production ranged from 197 233 to 257 602, although Lavigne himself felt that the actual number was less than 200 000. The adult population was generally held to be between four and five times the number of pups born, so it was possible that there were fewer than one million harp seals in the northwest Atlantic. Lavigne concluded that "if 1975 management policies were continued, production would continue to decline, and the population would be severely threatened before the end of the 20th century."

Lavigne made the findings available to the public via an article in the January 1976 issue of *National Geographic*.[61] IFAW seized upon the news even though, to that point, its opposition to the seal hunt was based primarily on ethical grounds. Moreover, ICNAF had reduced the harp seal TAC for 1976 to 127 000 (30 000 to landsmen, 52 333 to Canadian vessels, 44 667 to Norwegian vessels), a move that Lavigne himself called "a step in the right direction."[62] ICNAF also set a TAC of 15 000 for the much-neglected hood seal.

During the 1976 seal hunt, protestors converged on St. Anthony, Newfoundland, a small town near the Front. In addition to the media, Brian Davies brought along a group of American airline stewardesses who, he hoped, would spread the anti-sealing message upon their return to the United States.[63] He also thought that their presence on the ice would debunk what he perceived as the myth that sealing was a dangerous vocation. Davies was charged with violating Canadian sealing regulations by flying a helicopter within 2000 feet of a seal herd and by landing within half a mile of a seal herd. He was later acquitted because the offences occurred outside Canada's 12-mile territorial sea (the 12-mile limit had been proclaimed in 1970).

In 1976 IFAW had to share the limelight with a group that was new to seal-hunt protest. The Greenpeace Foundation had been formed in British Columbia in 1971 to protest American underground nuclear tests at Amchitka Island. It subsequently added to its reputation by attempting to interfere with Russian and Japanese whaling in the Pacific Ocean. At first nobody quite knew why Greenpeace was opposed to the seal hunt. Reports reached Newfoundland that the Greenpeacers intended to spray green crosses on the backs of the seals, supposedly because the sealers were "very much fundamentalist when it comes to religion" and would be reluctant to club seals so marked.[64] The 13 Greenpeace members who made the trip to Newfoundland abandoned this tactic after attending a public meeting at St. Anthony on 9 March.[65] They professed sympathy for local landsmen, pointing to the large profits that the sealing companies made from their labour, and emphasized the threat that large Canadian and especially Norwegian sealing vessels posed to their livelihood. Greenpeace gave no indication of how support for landsmen was compatible with its overall opposition to the seal hunt.

Canada assumed full responsibility for management of the seal hunt upon adopting a 200-mile fishing zone on 1 January 1977. Because of the expertise that had been built up within ICNAF, Canadian management decisions were based on recommendations from ICNAF (succeeded by the Northwest Atlantic Fisheries Organization in 1979) and COSS. One of Canada's first actions was to limit the number of adult seals taken by large vessels to five percent of their total catch. Again, this measure was intended to place greater emphasis on pups and immatures which, because of their higher natural mortality rates, could sustain higher kill levels. Most attention, however, focussed on the 1977 TACs for the gulf and Front, 160 000 harp seals (152 000 of them whitecoats) and 15 000 hoods. An additional 10 000 harps were allotted to native hunters in the Canadian Arctic. The new harp

TAC was historically important for two reasons. The allotment to landsmen was raised to 63 000 from its previous level of 30 000, where it had stood since 1972. Thus, for the first time, the landsman quota was higher than the quota for the Canadian fleet (62 000), reflecting a shift in the balance of power within the industry. The increased landsman allotment was achieved at the expense of the Norwegian fleet, whose allotment was set at 35 000 harps, down from the 60 000 level it had enjoyed from 1972 to 1975. There were six Canadian vessels to Norway's five in 1977, marking the first time since the Second World War that Canada had the larger fleet. After Canada declared the 200-mile fishing zone, the Newfoundland-based fleet entered a brief period of growth.

By this time, ICNAF scientists used not only aerial survey results but also an assortment of elaborate statistical models in setting the TAC. These models incorporated estimates of the crucial factors of mortality rates and reproduction potential. Disagreement over mortality rates had forced the lowering of the harp seal TAC to 127 000 in 1976, so ICNAF scientists devoted particular attention to that question, concluding that the actual mortality rates were considerably lower than many scientists, including Lavigne, had previously believed. In fact, Lavigne's model (the "Guelph model") contained "much higher" rates than any other model.[66] Thus ICNAF concluded that the harp seal population was increasing.

Not all scientists concurred with the ICNAF findings. Staffan Soederberg and Lennart Almkvist, of the Swedish Museum of Natural History, charged that the 1977 whitecoat quota was too high.[67] Specifically, they calculated that the population of immature seals would suffer a net loss of 17 393 in 1977, which reflected whitecoat overexploitation. Their conclusion was reinforced by ICNAF's own data showing that the mean age of sexual maturity among female harp seals had fallen to 3.8 years from 4. Soederberg and Almkvist maintained that the females were giving birth earlier in order to stabilize a declining population. If, as ICNAF stated, the population was actually increasing, the age of sexual maturity would also be increasing, or at least stabilizing.

Greenpeace director Paul Watson grabbed the headlines on the first day of the hunt, 15 March, by handcuffing himself to a whipline used to haul pelts aboard the *Martin Karlsen*. Prior to that, Watson and two other Greenpeacers had thrown pelts and clubs into the water, and Watson had even lain down on the ice in front of a sealing ship, forcing it to stop. No charges were laid and, surprisingly, Watson was taken aboard the *Martin Karlsen* and given dry clothes after getting dunked during the whipline incident.[68] Brian Davies continued his tactic of bringing women to witness the hunt, this time replacing airline stewardesses with actress Yvette Mimieux. As in 1976, Davies violated sealing regulations by flying his helicopter too close to the seals and also by landing too close to them. The difference this time was that the offence occurred within Canada's 200-mile fishing zone. Davies was later placed on probation and forbidden to fly over the seal herd during the hunt. This prompted accusations from the protest camp that the sealing regulations were designed not so much to protect the seals as to shield the sealers from the protestors.

IFAW and Greenpeace were joined in 1977 by millionaire Swiss conservationist Franz Weber, head of the appropriately named Franz Weber Foundation.[69] Weber offered to pay the Canadian government $2.5 million to stop the hunt, but he is best remembered for his proposal to establish an artificial-fur factory in Newfoundland to employ displaced sealers. Emulating the proven publicity tactics of IFAW, Weber brought with him to Newfoundland French actrress Brigitte Bardot, who stirred up the requisite anti-sealing sentiment in the French media.

The Counter-Protest Movement

Although international protest was aimed at Canada in general, Newfoundlanders understandably felt it more keenly than other Canadians since in 1976 they made up 86 per cent of all sealers in eastern Canada.[70] So it is hardly surprising that a counter-protest movement began to emerge on the island. The first sign of counter-protest appeared in 1976, when a group called the Concerned Citizens of St. Anthony organized a public meeting to confront Greenpeace.[71] The Trinity Bay Society for the Retention of Our Sealing Industry was formed in 1977 and later that year merged with the Concerned Citizens of St. Anthony to form the Progressive Rights Organization.[72] On 1 April 1977 the Newfoundland provincial government joined the fray, announcing its intention to launch an international campaign in defence of the hunt.

Why did Newfoundlanders wait until 1976 before launching a counter-protest movement? For one thing, the protest movement itself did not begin to mount significant strength until the mid-sixties. At that time, it focussed almost exclusively on the hunt in the Gulf of St. Lawrence. Newfoundland sealers at the Front and landsmen along the northeast coast were able to maintain a sense of detachment from the controversy. It was not until the Canadian government banned large vessels from the gulf in 1972 that attention shifted to the Front, and not until the release of Lavigne's findings in

January 1976 that the protest gained new momentum. As well, the protest movement had made no visible inroads into the economic well-being of the industry until 1976. Prices for seal pelts rose almost without interruption from $6.27 per pelt in 1968 to $20 per pelt in 1975.[73] Significantly, the price fell to $15.95 per pelt in 1976, coinciding with the rejuvenation of the protest movement. Industry officials claimed that the decline in prices after 1975 was due to a combination of adverse publicity and a shift in consumer tastes to longer-haired animals.[74] Some went so far as to lay all the blame on market whims, discounting any influence by the protestors.[75] However, negative publicity was a major reason for the fashion market's move away from seal. Finally, prior to 1977 Canada had suffered little official censure on account of the seal hunt. The United States Congress had passed the Marine Mammal Protection Act in 1972, banning the importation of products derived from nursing marine mammals under eight months of age, but by that date the American market for seal products was insignificant. However, on 22 March 1977, the United States House of Representatives passed a resolution by Congressman Leo Ryan of California condemning the hunt on humanitarian and conservationist grounds. It was this resolution that prompted the Newfoundland government to mount a counter-campaign. While the Progressive Rights Organization served largely to rally Newfoundlanders, the provincial government directed its campaign at the same national and international audiences that the protestors drew their support from.[76] In 1978 the government's team visited New York, Washington, Chicago, Toronto, Montreal, San Francisco, Vancouver, Winnipeg, London, and Frankfurt.

Pro-sealing groups use several arguments to defend the hunt, the most common being the hunt's economic value. The primary (hunting) and secondary (processing) sectors of the industry were responsible for injecting $5.5 million into the economy of Atlantic Canada in 1976.[77] Each of the region's 4030 sealers had an average of 3.5 dependents, so money earned from sealing directly benefited over 14 000 people. Most sealers lived in areas where employment was seasonal, determined by the length of the fishing season. During the winter months, these areas experienced chronic unemployment. Since 79.1 per cent of all Newfoundland sealers in 1976 possessed grade-nine education or less, they had limited occupational mobility even when other jobs were available. For some, sealing supplied one-third of their annual income.[78]

The hunt's supporters emphasize its place in Newfoundland's cultural heritage. Sealing has been a part of the Newfoundlander's life since the 18th century and was a key factor in settling the northeast coast. Traditional outport living in that region revolves around sealing in the spring, fishing in the summer and fall, and hunting wild birds and animals in the fall and winter. It is a pattern not all that different from the lives of the area's pre-contact native peoples. Playwright Michael Cook has charged that the seal-hunt protest movement is "in direct descent from those bureaucratic stupidities which interfered with native people's hunting rights, and sought to impose the values of an alien and indifferent culture upon that of others."[79] Outport society was and still is largely self-contained, as depicted in a Newfoundland song:

I'se the b'y that builds the boat,
I'se the b'y that sails her,
I'se the b'y that catches the fish,
And brings 'em home to Lizer.

According to this argument, the protest movement, born of an urban culture, fails to recognize sealing as an element of traditional Newfoundland culture.

The hunt's defenders insist that the seals are in no danger of extinction. Government scientists believe that the northwest Atlantic harp seal population is increasing and has been since 1972–73.[80] The harp seal TAC is routinely set below the maximum sustainable yield; that is, the maximum catch level at which the population will remain stable. For example, the TAC for 1978 was 180 000 seals, while the maximum sustainable yield was estimated at 220 000.[81] The protesters dismissed such figures on the grounds that government scientists cannot be trusted to be objective, an argument that failed to take into account how, in the 1950s, government scientists were the first to draw attention to declining seal populations.

Another popular pro-sealing argument is that if the hunt were stopped, the seals would threaten cod stocks and would eventually have to be culled. The problem, allegedly, is that a totally protected seal population would decimate caplin stocks, on which cod also feed. Of all the explanations used to justify the seal hunt, this is the most questionable one. There is at present insufficient data on the diet of harp and hood seals to support such a conclusion. David Sergeant has speculated that caplin constitute 25 per cent of all food harp seals consume, at which rate caplin stocks are not threatened.[82] The harp seal stock will not increase indefinitely, since any increase will be limited by the availability of food. And, as David Lavigne pointed out in 1978, "when fish stocks were more abundant than they now are, for example in the post-war years, no one suggested that seals were limiting the availability of fish, although the population of seals was two or three times its present size."[83] Indeed, one of the few areas where western Atlantic cod stocks are rebuilding is off the coasts of Labrador and northeastern Newfoundland, right along the migration routes of the harp and hood seals.

Supporters of the seal hunt also contend that it is humane. The issue has received considerable attention over the years, prompting one person to suggest that the seals are "more in danger of being studied to death than clubbed to death."[84] Numerous organizations have sent observers to the hunt at the invitation of the Canadian government, among them the Canadian Audubon Society, Ontario Humane Society, International Society for the Protection of Animals, and Canadian Federation of Humane Societies. In 1969 Dr. Keith Ronald of the Canadian Federation of Humane Societies examined 509 seal carcasses, of which all but one had crushed skulls, translating into a 99.8-per-cent successful kill rate. Dr. Lars Karstad, consultant to the New Brunswick SPCA, had found a 96.7-per-cent success rate in 1968. Either rate is comparable to that found in commercial abattoirs. Successive studies have shown that the club and Norwegian hakapik, adopted by Canadian sealers in 1976, are the most effective killing devices. The most extensive study was conducted for the United States government between 1969 and 1972 in an effort to discover an alternative to the clubbing used on the Pribilof Islands. Consultants included the American Academy of Sciences, Panel of Euthanasia of the American Veterinary Medical Association, Humane Society of the United States, Virginia Mason Research Centre, and Battelle Columbus Laboratories. They found that clubbing, followed by bleeding, was superior to other methods such as gunshot, acoustical shock, gas, poison, and electrocution. Against this wall of scientific evidence, the protestors counter that it is impossible for fisheries officers to supervise the sealers because they operate over too wide an area.[85] And, they reason, if sealers are not supervised, there is no guarantee they will use humane techniques.

Last, but by no means least, defenders of the hunt charge that protest groups are only in it for the money. One critic even called Brian Davies "the wealthiest sealer in Canada."[86] Davies himself has a ready response to that line of argument:

> Because you are working in this field you shouldn't have to take less money than somebody working in another field.... I think one of the critical weaknesses in the animal-welfare movement is that we're not paying the kinds of salaries that will attract well-paid professionals. And I consider myself a professional.[87]

The Protest Movement, 1977–1984

The protestors were beginning to suffer credibility problems. Brian Davies appeared at a London press conference on 17 February 1978 in the company of Raymond Elliott, a former Newfoundland sealer who claimed that half the seals taken were skinned alive.[88] It later transpired that Elliott had had his sealer's licence revoked in 1968 for improperly killing seals.[89] Since then, he had purchased seal meat from returning sealers at St. John's, selling it from his truck in and around the city. Elliott's brother felt he had joined forces with Davies because he stood to lose his vendor's licence owing to new health regulations.

IFAW had also run afoul of the taxman. Since its inception in 1969, IFAW had been a registered charitable organization and as such paid no income tax. Its revenues were substantial, having jumped 24 per cent to $1 020 828 in 1976, and a further 24 per cent to $1 267 826 in 1977.[90] The problem, as Revenue Canada saw it, was that IFAW was using that money for political purposes, that is, trying to stop the seal hunt. It therefore revoked the fund's tax-exempt status in 1977. The next year, the fund moved its headquarters from Fredericton to Yarmouth Port, Massachusetts. Under United States tax law, IFAW is regarded as a charitable organization and, accordingly, pays no income tax.

The Greenpeace camp was having its troubles too. Following the 1977 protest campaign, the foundation's board of directors expelled Paul Watson, allegedly because he had overrun the budget by $25 000.[91] Watson, who then formed the Earthforce Environmental Society, responded with his own accusations against Greenpeace.[92] He claimed Greenpeace was only interested in turning a profit and had gotten involved in the seal-hunt protest because it generated more revenue than any of the foundation's other activities. Watson contended that the many Californians coming to Newfoundland with Greenpeace's 1978 contingent was proof that their protest was out of place. Instead, he argued, they should be at home, protesting the killing of dolphins and sea turtles in Californian waters. Watson informed the St. John's *Evening Telegram* that "The extent of their knowledge about Newfoundland is that they have been briefed that 'Newfies' speak funny English. In fact, Newfie jokes are the rage in the Greenpeace offices down the Pacific coast."[93] Watson and Greenpeace president Dr. Patrick Moore bickered openly over Moore's salary, and Watson challenged Moore's motives. Moore, he said,

was against the seal campaign in the beginning because he thought
Greenpeace would be wasting money. He since discovered that
seals are in fact money-makers and for this reason he is opposed
to the seal hunt. This same man ... told me last year ... "I don't care
if they kill all the bloody seals just so long as they spell our name
(Greenpeace) right."[94]

The feud between Watson and Greenpeace reflected ill on both parties and raised serious questions about their integrity.

Despite these problems and the emergence of counter-protest in Newfoundland, the protest campaign suffered no noticeable damage. Paul Watson's Earthforce Environmental Society sent no representatives to Newfoundland to protest the 1978 hunt because, as Watson explained, "It looks like just another media circus shaping up and I don't want to get involved."[95] The abdication of Earthforce, together with the court decision against Brian Davies, left Greenpeace as the only protest group at the 1978 hunt. The protest began at Halifax on 27 February when 12 Greenpeacers in motorized rubber dinghies tried unsuccessfully to prevent the Karlsen-owned *Arctic Endeavour* and *Martin Karlsen* from leaving port. A second contingent arrived in Newfoundland on 7 March. It included, in addition to Dr. Moore, actresses Pamela Sue Martin and Monique van der Ven, and American Congressmen Leo Ryan of California (who was killed later that year in Jonestown, Guyana) and James Jeffords of Vermont. Moore was arrested twice, first for causing a disturbance by refusing to leave a temporary federal fisheries office at St. Anthony and later for interfering with the hunt by sitting on a whitecoat to prevent sealers from killing it. The first charge was subsequently dismissed, but Moore was fined $200 for interfering with the hunt.

Greenpeace's 1978 campaign was particularly noteworthy because, for the first time, the organization offered a comprehensive account of its objections to the hunt. It did so via a full-page ad in the St. John's *Evening Telegram*.[96] To counter charges that Greenpeace members were ignoring issues in their own part of the world, the ad drew attention to Greenpeace's opposition to the movement of oil tankers along the British Columbia coastline, nuclear testing in Alaska, pesticide dumping in British Columbia's Fraser Valley, and whaling in the Pacific Ocean. It then challenged a number of so-called myths used to defend the hunt. Chief among these was the belief that the hunt was of economic importance to Newfoundland. Quoting figures from a 1976 federal government survey of the hunt,[97] the ad noted that the value attributed to the hunt, $5.5 million, amounted to less than 0.2

per cent of Newfoundland's gross provincial income. And of that only $702 000, or 12 per cent, went to landsmen, the bulk going to the shipowners and Norwegian firms that processed the seal pelts into finished products. By comparing the industry's value to the gross provincial income, Greenpeace distorted its regional importance, a point that the author of the 1976 survey had taken pains to emphasize. In those parts of Newfoundland where seals were hunted, principally along the northeast coast, it frequently contributed one-third of the sealers' annual income. The $702000 attributed to landsmen was also misleading, in that Greenpeace did not consider longliner crews to be landsmen, which they were. (Longliners are decked boats used in the cod fishery. Their name comes from the long, baited lines they carry.) The value added by longliner crews was $1 202 474, which, combined with the $702 000, accounted for 34.6 per cent of the $5.5 million the industry generated. There was truth in Greenpeace's argument that the shipowners and foreign fur processors made most of the profit. Foreign control was indeed a historical weakness of the industry.

The second "myth" the ad challenged was that the seal herds were well protected by federal regulations and in no danger of extinction. The ad referred to a new ultraviolet census conducted by Dr. David Lavigne (actually by the University of Guelph, COSS, and Environment Canada) in 1977 estimating that year's whitecoat production at approximately 250 000, of which some 130 000 (actually 155 000) were killed by the sealers.[98] This was supposed to indicate that the herd was doomed to extinction. Yet 250 000 pups represented a 25-per-cent increase over Lavigne's 1975 estimate of 200 000. Moreover, the 1977 census suggested a maximum possible pup production of 300 000, an increase of approximately 42 000 over the 1975 level. In other words, the new census indicated that the herd was growing, not shrinking.

The ad's representation of the humane-killing issue was similarly deceptive. It cited a 1977 report prepared by University of Ottawa veterinary pathologist Dr. Harry Rowsell for the Canadian Federation of Humane Societies and COSS. Rowsell was said to have found that, of 76 seal pups examined, "unconsciousness was not instantaneous in 10 animals, or 14 per cent." Killing methods used in abattoirs, the ad claimed, were "much more reliable." Aside from the minor point that 10 of 76 is 13.2 per cent, the ad completely misrepresented Rowsell's findings, as Rowsell himself later retorted in a letter to the editor of the *Evening Telegram*.[99] Rowsell noted that the ad was designed to create the impression that seal pups were being skinned alive. But, he pointed out, the figures Greenpeace quoted related only to the landsman operation, in which beaters, not whitecoats, were shot by longliner crews. Moreover, "all seals not unconscious when brought on board the longliner were made so before pelting began. Nowhere did I suggest that any of these animals were conscious at the time of skinning." Rowsell also disputed the assertion that slaughterhouse killing methods were more reliable. He cited a recent European Economic Community (EEC) study that had found that "15 percent or more of the livestock were still conscious" when the animals were bled.

IFAW's absence from Newfoundland during the 1978 hunt created a false sense that the protest movement was faltering. Actually it was gaining momentum. IFAW was concentrating increasingly on arousing public opinion in Europe, especially Great Britain, against the hunt. In the long run, this approach would prove more effective than direct action on the ice floes. Late in 1978 the protest movement gained a new member in the form of the New York–based Fund for Animals. Founded in 1967, the fund was led by millionaire Cleveland Amory and included among its members Paul Watson, whose Earthforce Environmental Society had proved short-lived. In November 1978 Amory announced he was organizing an American tourist boycott of Canada as a pressure tactic to stop the seal hunt. By January 1979 the fund had distributed two million letters throughout the United States, with another million still to be mailed.[100] During the 1979 seal hunt the fund tried to disrupt the hunt in the gulf, which had been re-opened to a single large vessel in 1978 and to two in 1979. Together with the Royal Society for the Prevention of Cruelty to Animals, the fund purchased a British deep-sea trawler, renamed the *Sea Shepherd*, and sailed into the whelping ice. On 9 March eight *Sea Shepherd* crewmembers, including Paul Watson, were arrested for spraying red organic dye on more than 200 whitecoats. Watson was later fined $8300, sentenced to jail for 15 months, and placed on three years' probation, during which time he was forbidden to take part in any seal-hunt protest in Atlantic Canada. He served ten days before his release pending an appeal of the conviction.

Like IFAW, Greenpeace was expanding its protest effort on an international scale. In 1979, Greenpeace members showed up in Norway, where they chained themselves to the railings and masts of Norway's four sealing ships prior to their departure for the Front.[101] Twelve Greenpeacers were arrested in St. John's on Sunday, 4 March, during an ecumenical service and send-off for the sealers at the waterfront, an old custom that the Progressive Rights Organization had revived in 1978. As a crowd of five thousand looked on, the protestors used a combination of tactics to hinder the five ships' departure. Harbour police hauled one protestor out of the water before he could chain himself to a vessel's rudder. A few chained themselves to the vessel's railings, and others buzzed around the harbour in motorized rubber dinghies. Three protestors were later fined $100 each for obstructing navigation. The other nine were acquitted of charges of creating a disturbance during a religious ceremony and of causing mischief in relation to private property.

Greenpeace's official position on the seal hunt underwent a major change in 1979, a change that raised still more questions about the organization's credibility. At a St. John's press conference held on 2 March, Patrick Moore announced that Greenpeace was now opposed to the landsman hunt, which it had supported since 1976.[102] Greenpeace, he said, had been unaware that landsmen operated in boats up to 65 feet long, that is, longliners. Confusion about the status of longliner crews had been evident in their newspaper ad of the previous March. According to Moore, the confusion stemmed from misinformation supplied by federal officials. Thus, an organization that routinely rejected the government's seal population figures, its claims of humane killing, and anything else it said about the seal hunt, now expected the world to believe that it did not know what a landsman was because it had relied on incorrect information supplied by that same government.

After the episode at the St. John's waterfront, the action shifted to St. Anthony. The Greenpeace protest effort there had been entrusted to the foundation's Boston office. Owing to Moore's conviction in 1978, Greenpeace was restricted to a single permit to observe the hunt. Fisheries officers revoked the permit after Greenpeace co-ordinator Ed Chavies sprayed green dye on some whitecoats during his one trip to the ice on 14 March. The Greenpeacers could do little without permits, so they quickly left for the United States. The only other familiar figure among the protestors in 1979 was Franz Weber, who spent most of his time on the Quebec side of the Strait of Belle Isle. Without Brigitte Bardot he was not much of an attraction, and the media largely ignored him.

In the early 1980s Greenpeace and IFAW essentially maintained their respective approaches to seal-hunt protest, Greenpeace stressing direct intervention and IFAW courting international public opinion with a campaign that incorporated the latest marketing techniques. Paul Watson also remained a persistent critic of the hunt. There were others in the movement, but Greenpeace, IFAW, and Paul Watson in his many guises remained the leaders. In 1980, five women from Greenpeace's Boston office posed as members of a New York tour group that came to the Magdalen Islands to view harp seals. While on the ice, the Greenpeacers slipped away from tour guides and sprayed green dye on a few whitecoats. In a press release following the incident, Ed Chavies exulted that "the Newfoundland seal hunt is supposed to be one of the most closely regulated hunts in the world, yet Greenpeace members were able to reach and leave the seals without being detected."[103] Chavies seems to have assumed — erroneously — that fisheries officers were in the habit of supervising tour groups.

In 1981 Greenpeace turned its attention to hood seals. Although the hood TAC had held steady at 15 000 since 1976, scientific information on hoods was lacking. This was in no small measure owing to the protest movement itself. By focussing on the harp seal, the protestors had forced government scientists to devote their attention to it. Critics have even said that the preoccupation with the harp seal has retarded fisheries research on more important species, including cod.[104] To draw attention to the hoods, Greenpeace members sailed their protest ship *Rainbow Warrior* into the Front during the 1981 hood hunt. On 25 March, fisheries officers seized the *Rainbow Warrior* and arrested two of its crew for attempting to spray green dye on some seals. They and the ship's captain were fined $2000 each and forbidden to enter the gulf or Front sealing regions for three years. The *Rainbow Warrior*, with a new crew on board, was present in the gulf in 1982, when three more Greenpeacers were arrested for spraying dye on whitecoats. They were fined $1500 each and ordered not to visit Atlantic Canada during the seal hunt for the next three years. A handful of Greenpeace members were present at Charlottetown, Prince Edward Island, prior to the 1983 hunt, but left before the hunt began. Greenpeace sent no protestors to the seal hunt in 1984, primarily because the hunt had been so drastically reduced that protest was superfluous.

Paul Watson re-entered the protest movement in 1981 as a member of the newly formed Sea Shepherd Conservation Society, the fourth environmental group he had belonged to in four years. Watson, together with two associates, planned to set off for the whelping ice from Charlottetown, Prince Edward Island, in a three-man kayak. They intended to spray seals in the gulf, then to pass through the Strait of Belle Isle and do the same thing at the Front. The kayak was white, and the protestors planned to wear white clothing in order to avoid aerial detection. They had to abandon their plan because of weather conditions, but did claim to have sprayed some seals off Prince Edward Island's north shore.

Absent in 1982, Watson was back in 1983 with his most dangerous scheme ever. For nearly two weeks, the Sea Shepherd Conservation Society's protest ship *Sea Shepherd II* hovered outside St. John's harbour. Watson announced that it would ram any sealing vessel that attempted to leave the harbour. The *Sea Shepherd II* was protected by a barbed wire fence around its railings, and was said to be equipped with a water cannon to deter potential boarding parties.[105] While it waited, Newfoundlanders mulled over policy statements by the Sea Shepherd Conservation Society. According to director Michael Ballin, oil lay under the seal nurseries, so the sealing industry and the Canadian government were "very systematically killing off the seals to get access to the oil."[106] Paul Watson believed that not only seals, but "all marine mammals face extinction by the end of the century unless severe action is taken."[107] The seal hunt, said Watson, was "grossly inhumane, barbaric and brutal and it is a threat to our civil rights and liberties." The society repeated the now-familiar offer to pay sealers not to hunt. Watson mentioned a figure of $1.2 million that, he felt, would "probably" be put up by the Fund for Animals.

By 1983, because of the success of the protest movement, the entire Canadian sealing "fleet" consisted of three ships. The Newfoundland fleet had shrunk from six ships in 1982 (its highest number since 1953) to a single vessel, the *Clayton M. Johnson*, owned by Johnson Combined Enterprises of St. John's. It sailed to the Front from Catalina, frustrating Watson's original plan. Undaunted, Watson took the *Sea Shepherd II* into the gulf, where the hunt had been late starting. On 27 March, police seized the vessel off northern Cape Breton and arrested 17 society members, two days after Watson disobeyed an order to head for Cap-aux-Meules, Magdalen Islands. Police had issued the order because the *Sea Shepherd II* had come within half a nautical mile of landsman operations. In order to board the protest ship, a special Royal Canadian Mounted Police assault force first had to lob tear-gas canisters onto the deck and crush a section of the barbed wire fence by a gangplank dropped from the Canadian Coast Guard icebreaker *Sir William Alexander*. The water cannon proved to be a fake. This time the courts did not go easy on Watson. He was fined $5000 and sentenced to 15 months in jail. After serving nine days of the sentence, Watson was released in January 1984 pending an appeal of his conviction. In February and March 1984 he was back in the news as a result of his efforts to disrupt the British Columbia government's wolf-kill program.

No opponent of the seal hunt has been more persistent or more successful than IFAW. Although Brian Davies was the first to carry the protest to the scene of the hunt, IFAW has emphasized a different strategy since Davies' 1977 conviction for breaking Canadian sealing regulations. It has not completely abandoned the direct approach — Davies was in Charlottetown briefly during the 1981 hunt, and that same year a helicopter chartered by IFAW was impounded after it flew too close to the seals. Also, angry Prince Edward Islanders protested the presence of IFAW members at Charlottetown in 1983, and in 1984 hunt supporters in the Magdalen Islands inflicted an estimated \$350 000 worth of damage on an IFAW helicopter. But the fund has been most effective in mobilizing international public opinion against the hunt.

From the earliest stages of his involvement in the protest movement, Brian Davies was aware of the importance of carrying the anti-sealing message to the international audience. His breakthrough occurred with the London *Daily Mirror* and *Paris-Match* coverage of the 1968 and 1969 hunts. Through the 1970s IFAW used standard techniques like billboard advertising and protest marches in European capitals. But the most ingenious device has been the mail-in campaign, by which members of the public are urged to write a letter or mail a postcard (usually pre-printed) to some target organization. IFAW focussed on the European Parliament, which by March 1982 had received three million written requests urging an EEC boycott of seal products.[108] The European Parliament responded on 11 March by recommending a ban on the importation of products from harp and hood seal pups.

The bureaucratic structure of the EEC required that its administrative arm, the European Commission, consider the parliament's recommendation. If the commission decided to draft a ban regulation, the draft would be returned to parliament for final approval. The European Commission therefore became the focus of intense lobbying efforts by IFAW and protest groups on the one hand, and the Canadian and Norwegian governments on the other. Canada argued that the European Parliament's recommendation was based on false information regarding seal populations and killing methods. Canada therefore offered to co-sponsor a study to determine if Canadian killing methods were any less humane than those used in European abattoirs and, further, to refer the population question to the International Council for the Exploration of the Seas (ICES). ICES had been formed in 1902 as a multinational body of experts to provide scientific advice to its 18 member nations on matters pertaining to marine resources. For unexplained reasons, the European Commission refused to co-sponsor a study of killing methods, but did agree to refer the population question to ICES.

On 11 October 1982 the European Commission announced that it was recommending a temporary import ban. The commission based its decision on an anti-pornography clause in the General Agreement on Tariffs and Trade (GATT) allowing trade restrictions to protect public morals. Canada's minister of Fisheries and Oceans, Pierre DeBané, was outraged: the ICES report was still pending; Greenland, an EEC member, was excluded from the ban (because native hunters there killed only adult seals, which they shot with rifles); and the GATT morality clause had been invoked after the

European Commission turned down Canada's offer of a joint study of killing methods. Canada had more reason to be upset in November when ICES submitted its report. The report concluded that harp seal pup production in the period 1977–80 likely ranged from 380 000 to 500 000 annually, and that the population of harp seals over one year old was in the range of 1.5 million to 2 million. These figures were higher than the ones on which Canada had based its seal-management program in the same period. They also indicated that the harp seal population was increasing in size. ICES found that there was insufficient data to permit firm estimates of the hood seal population. Canada responded by lowering the 1983 hood TAC to 12 000 from its previous level of 15 000. On 19 November 1982, despite the findings of ICES, the European Parliament endorsed a temporary import ban effective to 1 March 1983.

All hope that the temporary import ban would be lifted was dashed on 28 February when the EEC extended the temporary ban to 30 September 1983, intending to impose a two-year ban starting 1 October. A final decision on the two-year ban would await the outcome of more talks between Canada, Norway, and the European Commission. Since EEC member countries imported approximately 75 per cent of Canadian seal pelts, this news had staggering implications. In 1983 there were only two large vessels at the gulf (one each from Nova Scotia and Quebec) and only one at the Front (from Newfoundland); no Norwegian ships went to the ice that year. Only some 30 000 immature and adult seals were taken even though there was a market for 60 000 such pelts within the EEC.[109] The problem was that market uncertainty had affected prices. Carino Company paid only $13 per pelt in 1983, half of what it had paid in 1982.[110] Ironically, in April 1983 the Convention on International Trade in Endangered Species, adhered to by 81 nations, ruled that neither harp nor hood seals were endangered species.

Matters went from bad to worse for the industry when, on 1 October 1983, the EEC implemented its threatened two-year ban (since extended to 1989) on the importation of products derived from harp and hood seal pups. Yet anti-sealing groups were still dissatisfied. Led by IFAW, they launched a vigorous campaign to end all sealing in Canada. In November IFAW appealed to the British public to boycott Canadian fish products as a means of pressuring the Canadian government to outlaw the commercial seal hunt. Canadian fish exports to Great Britain in 1982 were worth approximately $80 million, accounting for nearly ten per cent of Canada's fish exports. Within three months IFAW had flooded four and a half million British households with pre-printed postcards to be mailed to major supermarket chains urging that they remove Canadian fish products from their stores. On

6 February Tesco, Britain's largest supermarket chain, announced that it was phasing out all Canadian fish products and would not purchase any more until the seal hunt had ended for good. Tesco, with 465 stores, was soon joined by Safeway Limited, which had 105 retail outlets. IFAW organizers warned that unless the Canadian government banned all sealing operations in 1984, they would shift their campaign to the United States, which accounted for 80 per cent of Canadian fish exports, worth $967 million in 1983.

The Canadian government refused to give in. Indeed, it had already announced that it would guarantee Canadian sealers 80 per cent of the "normal" price ($30) for seal pelts in 1984. IFAW therefore carried through on its threat, announcing in late February that it would distribute five million pre-printed postcards to American households. The postcards could, in turn, be mailed to any one of five American companies buying Canadian fish, including the MacDonald's and Burger King restaurant chains. IFAW organizer Dan Morast informed the press that this was merely the first step in a larger program designed by the fund's marketing advisors.[111]

The compound effects of the EEC boycott and the potential fish boycott in the United States further emaciated the Canadian sealing effort in 1984. No large vessels sailed for the hunt, which was left entirely to landsmen. The threat to the Canadian fishing industry stimulated debate over the hunt's future. James Morrow, senior vice-president of National Sea Products Limited, Nova Scotia's largest seafood processor, urged that the hunt be banned outright. Morrow foresaw massive layoffs if American fish buyers began cancelling orders. Both the Newfoundland and Nova Scotia provincial fisheries ministers conceded that a ban might be necessary if the American fish market became endangered. The Fisheries Council of Canada proposed a moratorium on hunting whitecoats, while the Canadian Sealers' Association (formed in 1982) argued that a moratorium should include bluebacks as well as whitecoats. The Progressive Conservative opposition in Ottawa also favoured a whitecoat moratorium, reasoning that a moratorium would remove the chief objection to the hunt, that is, the killing of seal pups.

The Landsman Hunt

The recent history of the Newfoundland seal hunt has been characterized by the increased importance of the landsman operation. Despite the vessel-based hunt's larger overall contribution to the Newfoundland economy in the 19th century, the landsman operation continued to be significant in the more northerly bays along the northeast coast. Men often walked or used open boats to take seals near shore but, as in the 18th century, netting remained the preferred method.[112] The landsman operation was most important in Notre Dame Bay, in particular at Twillingate, New World Island, and Fogo Island, which juts out into the seals' migration routes. Greenspond was a key centre in northern Bonavista Bay in 1845, but its landsman operation contracted steadily thereafter as the population shifted towards Cape Freels, probably to gain better access to seals.[113] The population of Greenspond declined by 40 per cent between 1845 and 1857, while the overall population of northern Bonavista Bay expanded by 111 per cent. In the same period the number of seal nets in Greenspond fell by 280, but rose by 323 in northern Bonavista Bay. The landsman operation became less important in all communities as the century progressed, although Twillingate still counted 602 nets in 1884. This trend was the result of overkilling by vessel-based sealers at the Front.[114]

The lack of official statistics makes it difficult to plot the course of the landsman operation in the first half of the 20th century. The few figures available reflect the extreme variability of the operation. In 1924, landsmen on the Great Northern Peninsula took over 40 000 seals, and an additional 5000 were caught at Twillingate.[115] Yet in 1931 the entire landsman catch amounted to only 3000 to 5000 seals.[116] Another major problem confronting landsmen was getting their pelts to market, since their operation was highly decentralized. At the turn of the century, landsmen in Bonavista Bay brought their pelts to local merchants, who sent the pelts to St. John's by schooner as soon as navigation was possible in the spring.[117] Later, Bowring Brothers employed collectors who sailed to the outports to pick up pelts from local merchants acting as middlemen.[118] St. John's firms were not averse to buying pelts from landsmen outside Newfoundland. In 1927 Job Brothers sent SS *Thetis* to collect 20 000 pelts taken by landsmen on the Magdalen Islands.[119]

The landsman operation grew after the Second World War, although its relative importance was perhaps exaggerated by the decline of the vessel-based hunt. In 1951 the landsman catch of 40 000 seals constituted 22 per cent of the vessel catch.[120] Sergeant has estimated that, for the decade as a whole, landsmen contributed approximately 25 per cent of the total catch.[121] Significantly, in 1954 the landsman catch of 38 429 seals exceeded the vessel catch by 9501.[122] This was no longer unusual by the 1960s.

There have been two key innovations in the landsman operation during the 20th century. The first, just before the First World War, was the advent of the marine engine, which increased the range and mobility of landsmen's small, open boats. But more important was the introduction of longliners. Powerful, decked fishing boats 35 to 65 feet long, longliners had been introduced to Newfoundland on an experimental basis in 1951 to catch fish for processing plants along the northeast coast.[123] By 1959 there were 23 longliners with an average crew of 4.7 men.[124] The smaller ones could stay out for "several days at a time" and the larger ones for a week or longer.[125] This dramatically increased not only the landsmen's range but also their catching potential. Landsmen in Newfoundland and Labrador took approximately 50 000 seals in 1962 (just over 80 per cent of the total catch), earning $250 000 for the pelts plus additional monies from the sale of carcasses and flippers. As there were "several thousand" landsmen by the early 1960s, their individual earnings were still lower than those of the 400 or so Newfoundlanders who served on the large vessels. Nevertheless, as the Newfoundland Fisheries Commission reported in 1963, the landsman operation came

> at a most opportune time. Woodwork is finished and fishing has not commenced. Any earnings from Seals must, therefore, in some way or another, help toward the general economy of fishermen and enable them to get fishing with less credit.[126]

The best data on the landsman operation are from the 1970s. According to a 1976 Department of Fisheries and the Environment survey, there were 3841 landsmen (including longliner crews) in Atlantic Canada. Since 86 per cent of all sealers were based in Newfoundland and Labrador, there would have been approximately 3330 landsmen in the province. By contrast, the entire Canadian large-vessel fleet employed only 189 sealers, most of whom were Newfoundlanders. Longliner crews accounted for 20.7 per cent of all landsmen; the remaining 79.3 per cent hunted near shore, using snowmobiles and open motorboats, or walking to reach the seals. Landsmen were responsible for 58 per cent of the total catch of 78 127 seals in 1975. Their prominence was recognized in the 1977 harp seal quota, which more than doubled to 63 000, marking the first time the landsman quota exceeded the vessel quota (62 000).

By the 1970s, nets were becoming less and less popular. Their number dropped rapidly from 574 in 1971 to only 65 in 1973, most of them in Labrador.[127] Two reasons are generally offered for the decline.[128] Because the nets were worked nearly all winter long, weather and ice conditions made the labour difficult; freezing sometimes made it impossible to haul the nets, or else they actually sank. This led to the next problem, for if a seal was en-

53 Longliners at Twillingate, 1978.
Decks Awash Magazine, Extension Service, Memorial University of Newfoundland, St. John's.

54 Longliner *Ocean Bride III*, Twillingate, 1978.
Decks Awash Magazine, Extension Service, Memorial University of Newfoundland, St. John's.

55 Seal being taken aboard the longliner *Green Bay Rover*, April 1973.
Photo by Fred Earle, Lewisporte, Newfoundland.

tangled in a net for too long, the ropes left marks on the pelt, reducing its value. Such damaged pelts were said to be "wormy."

Longlining had become even more popular by the 1970s, perhaps another factor in the sudden swing away from nets. Sealing licences were issued to 199 longliner crews in 1976. Revenues longliner crews earned that year ($1 204 220) were 1.7 times higher than those of all other landsmen ($705 045), proof of the longliner's greater productivity.[129] The average longliner crew of 4.7 men was unchanged from 1959, with crews ranging from 3 to 10 men.

The longlining season in northeastern Newfoundland began in January while some seals were still migrating southward. Ice conditions at that time of year normally restricted the hunters to day trips.[130] Peak hunting occurred in March and April as the seals returned northward, the looser ice permitting trips of around one week's duration. Longliner crews often ventured far from their home ports. Crews from Little Bay Islands in Notre Dame Bay went as far north as the Horse Islands or even Groais Island. Unlike the crews of larger vessels, longliner crews took very few whitecoats or bluebacks. Over 80 per cent of their catch consisted of adult harps, bed-lamers, or beaters. Accordingly, rifles were preferred to clubs. The best hunting was in the early afternoon while the seals sunned on the ice. In a five-man crew, two men were gunners, two hauled the seals aboard, and one steered the longliner. The gunners shot the seals as the longliner passed the herd. Then the boat was manoeuvred alongside the carcasses so they could be hauled aboard, although sometimes men struck out in the longliner's motorboat to collect the seals.

The seals were brought aboard round and bled immediately. After pelting, the skin was washed and anti-oxidant oil was applied to prevent the pelt from yellowing. Cooled pelts were stored in the longliner hold, which ideally was lined with plastic or some other suitable material to prevent the pelts from touching rusty metal parts. Because of a lack of proper storage facilities on shore, many landsmen buried their pelts in the snow until the spring. Such pelts were said to be "snowed."

With the extension of the provincial highways network in the 1960s, the local merchants who bought landsmen's pelts could truck them to a buyer/processor at St. John's, Halifax, or Dildo (the Carino Company plant opened at Dildo in 1969). In the mid-seventies some landsmen began selling directly to the processor, hiring truckers to transport their catch. Many landsmen subsequently transported their own pelts after finding that the truckers charged high rates and did not protect the pelts against rust. Actual practice varied according to the community. A similar situation obtained for flippers and carcasses. These were sold to local merchants or to one of the seal-meat-processing plants, or were trucked to the large St. John's market and sold directly to consumers.

Despite the increasing participation of longliners in the landsman hunt, the vast majority of landsmen (79.3 per cent in 1976) relied on other methods. Under the proper wind and ice conditions, the seals were brought close enough to shore that the hunters needed only to walk out to club whitecoats or shoot adult seals. Some hunters even used snowmobiles to reach the seals. However, landsmen more commonly operated from small (under 35 feet) open boats equipped with outboard motors. The more adventurous hunters took their motorboats as far as 25 miles offshore, but most stayed within 12 miles. Their season began as early as November in parts of Labrador and in January along the northeast coast of Newfoundland.

The one advantage that motorboat owners enjoyed over longliner owners was lower capital costs. Indeed, most longliners operated at a marginal loss. Many were not reinforced for ice navigation and hence were prone to ice damage. Likewise, cold weather frequently caused electronic equipment to break down. Such damage necessitated expensive repairs and contributed to the vessel's depreciation. So why did longliner owners bother to go sealing at all? A 1976 survey revealed that because of anticipated revenues from other fisheries following the seal hunt, the owners "appear content to return the majority of revenues to labour rather than to capital over the short period of their sealing operation." Thus, while longlining was not lucrative for the owners, it was worthwhile for the crew. The average net return to labour for longliner crews in 1976 was $871.39 per man, compared with $108.90 for all remaining landsmen. Significantly, one of the other fisheries pursued by longliner crews, the Labrador fishery, began to attract longliners from northeastern Newfoundland in the 1960s.[131] The longliner therefore constituted another link in the historic relationship between sealing and the Labrador fishery.

While not as remunerative as the vessel-based hunt, the landsman operation was of considerable economic importance. Depending on the community, it provided anywhere from one-tenth to one-third of a fisherman's annual salary. The future of the landsman operation, like the sealing industry in general, is threatened by the EEC import ban, although landsmen will probably always enjoy a local market for carcasses and flippers. Beyond that, however, lie other problems. Landsmen on the Great Northern Peninsula claim that by the time the seals reach them in the spring, landsmen from Notre Dame Bay have already taken most of the quota. Similarly, motorboat crews complain that longliner crews take most of the catch before the seals come close to shore. Also, the expansion of the landsman operation in the 1970s created a glut of seal meat that drove down prices. To address these and other issues, landsmen formed the Canadian Sealers' Association in 1982.[132] That organization's call for a moratorium on hunting whitecoats and bluebacks reflected its makeup, for such a moratorium would have little effect on the landsman operation.

The Vessel-Based Hunt Since 1939

Preparations

The shift to motor vessels during and after the Second World War increased the number of communities outside St. John's that sent vessels to the hunt, and that meant a wider distribution of the money spent readying the vessels as owners turned to their local merchants for services and provisions. However, the motor vessel fleet shrank dramatically during the 1950s, and St. John's resumed total dominance after Earle Freighting Services of Carbonear withdrew from the industry after the 1966 season. Although Johnson Combined Enterprises had its headquarters at Catalina until 1978, St. John's was always its practical base of operations.[1] It cost Johnson Combined Enterprises $250 000 to outfit and fuel its three ships in 1982, plus an additional $50 000 to $60 000 to buy new life rafts and repair radios and mechanical equipment.[2] This money all went to St. John's businesses.

The pattern of sealers converging on their departure points remained a familiar one down to the last years of the large-vessel era. Sealers going to St. John's were able to use the train until the late 1960s when Canadian National Railways phased out its rail passenger service in the province. Then the chief mode of transport became car or bus over the Trans-Canada Highway, completed in 1965. After confederation there was also some westward traffic to Halifax by those men going to sign up with the Nova Scotian fleet.

The movement of the sealers was imperceptible to most Newfoundlanders in this period, for it was no longer a migration involving thousands of people. Only 189 men served on the vessel fleet in 1976; a hundred years earlier there had been 3678. To the very end there were not enough berths to go around, but demand did slacken considerably after 1949 with confederation and its attendant social benefits.[3]

Vessels

The use of auxiliary schooners in the interwar period presaged the adoption of fully motorized vessels in the 1940s. The first motor vessels were minuscule in comparison with wooden walls: the average motor vessel of the immediate postwar period was built of wood and was around 100 gross tons.[4] However, as had sailing vessels and steamships, the motor vessels soon got bigger. In 1951 the 12 motor vessels of the Newfoundland fleet ranged from 170 to 750 gross tons.[5] But they were hardly awe-inspiring. One observer remarked in 1955 that the Canadian sealing fleet had "developed into a motley collection of motor vessels, re-converted wartime craft, small whaling ships and combination trawler-sealers."[6] The only impressive craft among them were based in Halifax. For example, the Scottish-built trawler-sealer *Theta* (224 net tons), owned by Christensen Canadian Enterprises, had a refrigerated hold and an all-welded reinforced steel hull.[7] Its electronic transmission allowed it to be steered from either the bridge or the crow's-nest. The 155-foot vessel contained shower and laundry facilities for its crew of 40 sealers, with individual cabins for the captain, mate, engineers, radio operators, and steward.

In 1961 the Newfoundland "fleet" consisted of a single ship, Bowring Brothers' *Algerine* (338 net tons), a former United States Navy ocean-going tug built in 1943.[8] The *Algerine* was practically new compared with Earle Freighting's *Kyle*, which went to the ice in the mid-1960s. Built at Newcastle-on-Tyne in 1913, the *Kyle* was a former government coastal steamer, affectionately known as "the last of the coalburners." The first genuinely modern Newfoundland sealing vessels of the postwar era did not appear until the early 1960s. These were the *Sir John Crosbie*, built in 1962, and the *Chesley A. Crosbie*, built in 1964, owned respectively by Newfoundland Engineering and Construction Company Limited and Chimo Shipping Limited, both subsidiaries of Crosbie and Company. These two were big steel ships of over 1000 net tons, built to withstand heavy ice.[9] Primarily intended for transporting cargo to arctic regions, they were easily adapted to sealing.

In 1951, 10 of the 12 motor vessels in the Newfoundland fleet were locally built.[10] Over the next three decades, however, nearly all sealing ships originated outside the province. The *Chesley A. Crosbie* and its sister ship were built by Port Weller Dry Docks Limited of Port Weller, Ontario. The *Algerine* was an American ship, but it was at least modified in St. John's, being converted from steam to diesel power.[11] The *Lady Johnson* (95 net tons) was indeed a Clarenville, Newfoundland, product, but the *Lady Johnson II* (273 net tons) had been built in Norway in 1932 and the *Clayton M. Johnson* (248 net tons), built in 1960, also came from Norway. Carino Company, which moved its base of operations from Nova Scotia to Newfoundland in the late 1970s, also purchased its vessels in Norway. Puddister Trading Company's *Polar Explorer*, originally the Karlsen-owned *Theron*, was built in Scotland in 1950.

Because the postwar fleet had so many fewer ships, the owners more easily found other uses for their vessels when they were not sealing, an economic necessity given the brevity of the sealing season. Actually, the time had long since passed when vessels were used primarily for sealing. The *Algerine* was chartered for four months of each year to the Canadian government for hydrographic surveys in the Arctic, and for the winter months to the Anglo-Newfoundland Development Company as an ice-breaker.[12] In 1977 the *Lady Johnson II* was chartered by Imperial Oil for research in Davis Strait, while Carino Company's *Arctic Explorer* was chartered by the Newfoundland Oceans Research and Development Corporation, which was studying ice conditions off Labrador.[13] Sealing ships were also chartered to the Department of Fisheries and Oceans for research and patrol duties.[14] In these ways the sealing companies maximized the use of their vessels.

Season Length

The 1920s saw signs of indifference to the sealing regulations, especially in connection with second trips.[15] That indifference became widespread during and after the Second World War, with many ships making two and sometimes even three trips.[16] Sealers also flouted the 13 March opening date and the prohibition against killing seals on Sunday (although Sunday killing became legal in 1941). Again, Norwegian competition was blamed.[17] Unrestricted competition was finally brought under some control in 1952 when Canada and Norway agreed to opening dates of 10 March for the Front and 5 March for the Gulf of St. Lawrence. (After 1949 the hunt was subject to Canadian regulation, and not to pre-confederation Newfoundland legisla-

56 SS *Kyle* in Halifax harbour, February 1963.
Maritime Museum of the Atlantic, Halifax, Neg. No. N-12,175.

57 MV *Lady Johnson II* at the Front, March 1982.
Canada, Department of Fisheries and Oceans, St. John's; W.D. Bowen photo.

tion.) However, the failure to set closing dates allowed indefinite continuation of the hunt, which placed particular strain on the breeding population. This problem was not addressed until 1961, when Canada and Norway agreed to a 5 May closing date for both sealing regions. Canadian sealing regulations up to 1970 continued to focus not so much on departure dates for the fleet, on which pre-confederation Newfoundland legislation had concentrated, as on the actual dates when killing began and ended. The Seal Protection Regulations of October 1964 established seasons of 12 March–30 April at the Front and 7 March–25 April in the gulf, with the gulf dates subject to change at the minister's discretion. Thus, for example, the opening date in the gulf was changed to 8 March for the 1968 season. That same year Canada and Norway agreed to a 22 March opening date at the Front for one year only, with the intention of reducing pressure on the whitecoat population (this action was repeated in 1970). The Front season was further shortened in 1968 when a 25 April closing date was adopted. After 1964, sealing ships commonly left St. John's around 8 March for the Front and a week earlier for the gulf. Ships headed for the gulf moved on to the Front as soon as the gulf quota (50 000 seals) was met, normally within three or four days of the opening date.

After 1970, closing dates were affected by the quota system. In 1971 the season began on 12 March in both regions and ended either on 24 April or when the quota was reached, whichever came first. These regulations were further refined for the Front in 1973 (large vessels having been banned from the gulf in 1972). The season for harp seals began on 12 March and for hood seals 20 March, while the season for both species ended on 24 April or when the quotas were reached. The opening dates fluctuated only slightly from this for the remainder of the decade and into the 1980s. Such annual variations were the result of on-the-spot inspections by federal fisheries officers, who took into account such factors as weather conditions and the birth date of the seals. Large vessels were re-admitted to the gulf region in 1978, but only if they were from Quebec or Nova Scotia. Thus Newfoundland vessels continued to sail to the Front only, generally leaving port during the first week of March. The only exception was in 1983 when, because of the EEC import ban on seal-pup products, the lone Newfoundland ship did not leave until much later in the month.

Navigation

Navigation was one area in which the sealing industry experienced tremendous gains during the modern era. After the *Newfoundland* disaster, all ships had to carry wireless. This improved communications within the fleet and enabled the captains to get up-to-date weather forecasts. The next major advance was the adoption of radar in the 1940s. An important advantage after 1949 was the availability of at least one Canadian Coast Guard icebreaker to free jammed ships. Nevertheless, the icebreaker could not be everywhere at once, and sealing ships had to be prepared to survive on their own. Unfortunately, many of the early motor vessels were not able to do this, and until the mid-fifties hardly a year went by without one of them being claimed by the ice. After quotas were adopted and competition subsequently eliminated, the more powerful sealing ships commonly went to the aid of weaker vessels that had become jammed.[18] The usual technique was to break the ice in a circle around the jammed ship.

Steel motor vessels not surprisingly fared much better than their wooden counterparts, as most were built with a view to ice navigation. Ideally the vessel had extra shell plating, intermediate frames between the main frames along the length of the ship, stronger-than-normal steel at the bow and stern, and a specially constructed icebreaker bow.[19] It was up to each captain to choose a route through the ice most suited to his ship. His choice depended on such factors as the nature of the ice and the tonnage, horsepower, and strength of his ship. For example, great horsepower was not necessarily an advantage if the ship did not also possess a strong bow.

The actual icebreaking methods were not all that different from those used for the wooden walls. If the ice could not be crushed by the weight of the ship, the captain resorted to butting. When this failed he would try to blast through, but in the motor-vessel era, dynamite replaced home-made bombs.

Crews

With the advent of the first motor vessels, crew size fell dramatically. The five motor vessels that participated in the 1945 hunt carried an average crew of only 23 men. Crews did grow as the small motor vessels were replaced by larger ones, but only to a fraction of prewar levels. The average rose to

58 MV *Arctic Endeavour* at the Front, March 1979.
Canada, Department of Fisheries and Oceans, St. John's; W.D. Bowen photo.

91 in 1950, then tailed off to roughly 60 over the next five years. Some of
the larger vessels that decade were Captain J.H. Blackmore's *Saint Addresse*
with 85 sealers; Bowring Brothers' *Algerine* with 70; and the *Arctic
Prowler*, owned by the Halifax firm Sea Traders Limited, with 85.[20] Without
these vessels the average would have been much lower. In 1957, for example, the *Glenwood*, *Bessie Marie*, and *Placentia* all had crews of under
30 men.

By the 1960s there were so few Newfoundland sealing vessels that to
speak of averages is misleading. In 1964 Earle Freighting's *Terra Nova* carried only 25 men, while another of the firm's ships, the *Kyle*, carried 100.[21]
In 1965 *Algerine* counted 60 sealers to 80 for the *Chesley A. Crosbie*.[22] By
1970 there were only two Newfoundland-owned ships, the *Chesley A. Crosbie* and the *Lady Johnson*, with crews of 65 and 18 sealers respectively.[23]

The existence of quotas after 1971 made a large crew an economic
liability. To avoid cutthroat competition for the available seals, Newfoundland and Nova Scotian shipowners made a gentlemen's agreement to
share the catch (9500 harps and 1000 hoods per ship in 1982).[24] With a finite
number of seals apportioned to each ship, an owner recruited only as many
men as were necessary to catch his vessel's allotment. Crews got noticeably

smaller after 1973. In 1974 Mayhaven Shipping Limited, a Halifax firm that recruited exclusively in Newfoundland, pared the crew of the *Arctic Explorer* by nearly 20 per cent — from 59 to 48.[25] Three years later, the *Arctic Endeavour* carried only 20 sealers.[26] In the 1980s, Johnson Combined Enterprises' *Lady Johnson II* and *Clayton M. Johnson* sailed with crews of 25 to 30 men each; a decade earlier the same ships would have carried 70 to 80 men.[27] The smallest crews of all were found on the vessels of Karlsen Shipping Company Limited, which, after quotas were imposed, recruited crews of around 18 men per ship.[28]

Canadian sealing regulations introduced in March 1976 stipulated that a "sealing group" (formerly a watch) was to consist of not less than four and not more than ten men, one of whom was to be designated group leader.[29] The group leaders on Karlsen ships were the first and second gunners, who were also the first and second mates. By the late 1970s more than two groups per ship were rare.

The motor vessel greatly reduced the sealers' shipboard duties. Without sails to manage and coal to fuss with, the men had a lot of free time while the ship journeyed to the ice. They were not completely idle: there were side sticks to build, whiplines to rig, tow ropes to splice, sheaths to whittle, and hakapiks to adjust. But the sealers were still left with spare time for card games, checkers, and story telling.

Non-sealers on most motor vessels included the bosun, cook, cook's helper ("cookee"), steward, first and second mates, oiler, and engineers, whose number varied according to the size of each ship's engine. The development of electronic gears for motor vessels made the position of barrelman or scunner redundant. There was only one barrel, located on the foremast, and the captain usually occupied it from early morning until dark. There he not only looked for seals and picked out his course, but also steered the ship through the ice.[30] This further reduced sealers' shipboard duties by doing away with bridgemasters and wheelmen.

In the early years of the motor-vessel era, northern Trinity Bay sent several vessels to the ice, employing many sealers from that area. Captain Morrissey Johnson has estimated that they numbered between 150 and 200 men.[31] If those figures are accurate, northern Trinity Bay would have displaced northern Bonavista Bay as the main source of labour (from 1945 to 1955 the annual average number of vessel-based sealers was 520). By the 1970s, sealing companies recruited from a wide area. The crew of the *Arctic Endeavour* in 1977 came from Torbay, Bonavista, Centreville, Wesleyville, Twillingate, and St. Anthony.[32] Johnson Combined Enterprises deliberately avoided recruiting too many men from any particular area in order to reduce the effects of a possible disaster on any one community.[33] Whether other firms followed this policy is not known.

Equipment

While the sealing industry had known technological change throughout its history, in one area change was slow to come. The basic equipment of the sealer in 1960 was the same as it had been in 1860: tow rope, sculping knife, and gaff. This was not altered until 1967. Humane societies' continued complaints that the gaff was an inhumane killing implement led to the banning of gaffs prior to the 1967 hunt. The argument was that the gaff hook was occasionally driven into live seals, causing unnecessary pain. All sealers were therefore obliged to use a club not less than 24 inches and not more than 30 inches long. The club, for at least half its length, could not be less than 2 inches in diameter.

The banning of gaffs angered Newfoundland sealers. They maintained that it was an effective tool if used properly. A second and more common argument was that the loss of the gaff adversely affected safety on the ice. As one sealer put it, "you go into the water with a club you might just as well be dead right away."[34] Before, if a sealer fell through the ice he could save himself by driving his gaff hook into solid ice and hauling himself out of the frigid water. When Newfoundland Premier Joey Smallwood visited the Front on board the *Chesley A. Crosbie* in 1971, he told the sealers he would try to bring back the gaff.[35] The issue was all the more aggravating because Norwegian sealers working alongside the Newfoundlanders used hakapiks, which closely resembled gaffs.

The Canadian government responded by appointing Dr. Harry Rowsell, advisor to COSS and the Canadian Federation of Humane Societies, to study the possible adoption of the hakapik. Rowsell ruled in favour of the hakapik, and in 1976, Newfoundland sealers took to the ice for the first time with hakapiks. The hakapik was defined in that year's sealing regulations:

> *"hakapik" means an implement made of iron having a slightly bent spike of not more than five and one-half inches in length on one side of a ferrule and a blunt projection not more than one-half inch in length on the opposite side of the ferrule, the whole to weigh not less than three-quarters of a pound and having a head securely attached to a wooden handle not less than forty-two inches or more*

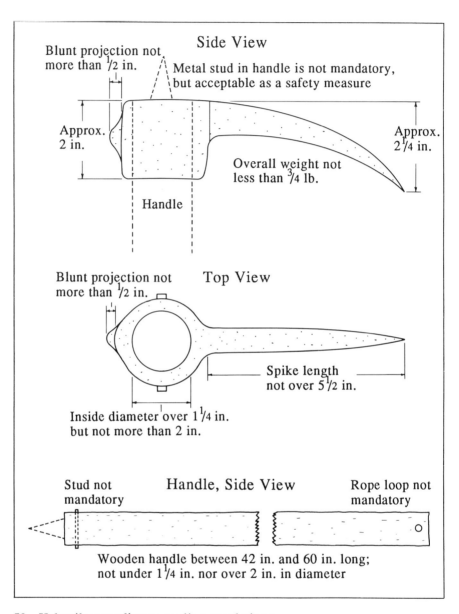

Side View

Blunt projection not more than $1/2$ in.

Metal stud in handle is not mandatory, but acceptable as a safety measure

Approx. 2 in.

Approx. $2 1/4$ in.

Overall weight not less than $3/4$ lb.

Handle

Top View

Blunt projection not more than $1/2$ in.

Spike length not over $5 1/2$ in.

Inside diameter over $1 1/4$ in. but not more than 2 in.

Handle, Side View

Stud not mandatory

Rope loop not mandatory

Wooden handle between 42 in. and 60 in. long; not under $1 1/4$ in. nor over 2 in. in diameter

59 Hakapik according to sealing regulations.
Drawing by Wayne Hughes.

*than sixty inches in length and with a diameter of not less than one
and one-quarter inches or more than two inches.*

The vessel owners supplied hakapiks to sealers.[36]

Substituting the hakapik for the gaff meant that a sealer's basic equipment was once more comparable to what it had been in the 19th century. From 1976 until the last trip in 1983, each man still carried a sculping knife (with either a curved or straight blade according to individual preference).[37] Sculping knives were sold from the vessel's stores, as were sharpening steels. A sealer purchased a knife and steel only when his old ones wore out. Both the steel and sculping knife in its wooden sheath were hung on the sealer's leather or rope belt, which was worn outside his clothing. Looped around the sealer's shoulder was the tow rope, its hook turned away from his body for safety's sake.

Clothing and footwear became much more practical after the Second World War. Leather boots were replaced by rubber ones (often snowmobile boots). Even though treated leather was fairly water resistant, rubber was superior. It kept more water out and retained more body heat. Once a leather boot had soaked up water and seal blood, it was useless as an insulator. The need for studs was lessened after the war because sealers did not have to haul pelts such long distances, motor vessel crews tending to work much closer to their vessels than steamer crews. Nevertheless, when extra traction was necessary, the sealer put on a pair of "creepers," spiked metal pieces that fitted over the soles of his boots.[38] A few diehards persisted in driving nails into the heels of their rubber boots. Instead of mittens, the modern sealer wore insulated rubber gloves. Most also wore oilskins over several layers of clothing that included coveralls. Goggles or sun glasses remained essential to protect against ice blindness. Brian Davies observed in 1966 that some sealers, instead of wearing goggles, simply daubed seal blood around their eyes.[39]

The Whitecoat Hunt

Prior to the Second World War, whitecoats accounted for nearly 90 per cent of the Newfoundland sealing fleet's total catch.[40] This reflected the industry's desire to maintain the breeding population, the soundest policy for management of the resource. Guns, used almost exclusively for killing beaters and adult seals, were restricted and sometimes banned outright in the interwar period. It is doubtful that these restrictions were closely observed, but this did not detract from the overwhelming emphasis on whitecoats. Only after the war did the catch of older seals increase (to 40 per cent of the northwest Atlantic seal catch by 1955) as a result of Norwegian hunting practices.

Anthropologist Guy Wright's eyewitness account of the 1979 seal hunt attests to the continuity of hunting methods, but also to change wrought by the protest movement.[41] A typical day at the ice began when one of the mates roused the men at 3:50 a.m. Before sitting down for breakfast at 5:00 a.m., they had to stow any pelts caught the previous day and left on deck to cool. After breakfast they gathered on deck at 5:45 a.m., going over the side around 6:00 a.m. Most vessels had side sticks, but some had only ladders or even car tires tied together.[42]

The men's first task once they were on the ice was to select a good collecting pan. When one was found, the men in the group spread out from the pan in search of seals. One effect of the drastic reductions in crew size was that the sealers now worked separately instead of in pairs.

The actual method of killing whitecoats reflected the new awareness of humane killing, a legacy of the protest movement. The sealer struck the whitecoat not twice but three times with the hammer end of his hakapik. The first blow usually caused a massive brain haemorrhage, killing the seal or at least rendering it "irreversibly unconscious." He then delivered two more blows to be certain. At this point some sealers checked the seal's eyeball for signs of reflex activity, and if they saw any, struck the seal until the reflex passed.

The seal was pelted in the established way, beginning with a cut along its underside from chin to tail that completely disembowelled it. The sealer then used his knife to separate the skin and fat from the carcass. Depending on the captain's preference, one or both front flippers were left on the pelt, with small holes cut in them for the tow rope. The small hind flippers, containing little meat, were always removed. A veteran sealer could pelt a whitecoat in about three minutes. The task required considerable skill, for even a slight nick reduced the pelt's value. Surprisingly, no special measures were taken to ensure that novice sealers developed the necessary expertise:

> *For the green hands, experience is the best teacher. A young rural Newfoundlander is generally expected to learn by imitating an experienced person who shows him a technique once or twice. Youngsters are not taught in the usual sense of formal training. And so Norman showed me how to skin a seal, and after doing so only twice, he expected me to have an adequate grasp of the technique.*

He did little explaining. When I badgered him with questions about the third seal, he made it plain that I was annoying him and that he wanted to get on with his work.[43]

If this practice was widespread, it must have caused considerable financial loss over the years as young sealers nicked and hacked their way to skill.

After the whitecoats were pelted, they were towed to the collecting pans, each tow consisting of three to five pelts. In the 1970s there were seldom more than 100 to 120 pelts per pan. Each pan was marked with a flag unless two pans were fairly close together, when only one pan needed to be marked.

Unlike in the steamer era, sealers were not routinely left out on the ice for the entire working day. The ship no longer had to move as far from the men because 20 men were incapable of covering as large an area as 200. Also, following the imposition of restrictions on overnight panning in 1964, no pelts had to be picked up from the previous day except when there had been bad weather. Accordingly, the captain often used a megaphone to summon the men aboard at noon for a quick, 20-minute lunch. If the captain knew he could not pick his men up at noon, the cook put lunch ingredients out at breakfast and the men carried their meal onto the ice, leaving their lunchboxes on the collecting pans while they hunted seals. The sealers concluded their work on the ice around 6:30 p.m.

During the day and into the night, men on the vessel were busy winching the pelts aboard by means of wire-cable whiplines. Each vessel had two or three whiplines of different lengths, and all could be used simultaneously.[44] The largest whipline on Karlsen ships was approximately five thousand feet long, allowing the ship to pick up pelts a considerable distance away.[45] Small groups of about 15 pelts were attached to the whipline at regular intervals.

As soon as the sealers were back on board the ship, they washed and ate supper, then put their oilskins back on and made their way to the deck for the evening's work, each man choosing the task he was most comfortable with. The pelts were hosed down as soon as possible after being winched aboard. After this, the front flippers were cut off and piled in buckets. The buckets in turn were emptied into wooden bins on deck where the flippers remained until they froze. Once frozen, they were removed to the hold. The pelts were stacked on the deck blubber up, fur down, to drain and cool. This ordinarily took until the following morning or evening. One postwar innovation, borrowed from the Norwegians, was the application of anti-oxidant oil to the fur side to guard against yellowing. On Newfoundland ships, the anti-yellow, as it was called, was applied immediately after the pelts were hosed

60 Sealer at the Front, 1979, with his oilskins, hakapik, sculping knife, steel, and tow rope.
Canada, Department of Fisheries and Oceans, St. John's; W.D. Bowen photo.

61 A pan flag from the *Lady Johnson II*, 1981.
Canada, Department of Fisheries and Oceans, St. John's; W.D. Bowen photo.

62 Pelts on a whipline, the Front, 1979.
Canada, Department of Fisheries and Oceans, St. John's; W.D. Bowen photo.

63 Deck scene, *Lady Johnson II*, the Front, March 1979.
Canada, Department of Fisheries and Oceans, St. John's; W.D. Bowen photo.

down, whereas on Karlsen ships it was not applied until the day after the pelts were cleaned. While some of the men were removing flippers and cleaning or stacking pelts, others were busy winching aboard pelts that the ship had failed to pick up during the day, usually because of problems with the ice. Night work might last until 1:00 a.m.

After less than three hours of sleep the men were expected on deck again to move cooled pelts into the hold. Each man took two pelts at a time to the hatch and tossed them below to other men who stacked them in the wooden pounds, removing rock ballast to make room for the pelts. Although most steel ships possessed refrigerated holds, some did not. In this respect the Norwegians were far ahead of the Canadians. They separated the fat from the skin while still at sea, salting the skins and storing the fat in tanks. On Newfoundland vessels the pelts, consisting of skin and attached fat, were stored with a bit of crushed ice or a sprinkle of salt between each layer.

64 Sealers killing adult harp seals, the Front, 1979.
Canada, Department of Fisheries and Oceans, St. John's; W.D. Bowen photo.

The Hood Hunt

The seal hunt of recent times was not structured along the old lines of a whitecoat hunt, followed by a beater hunt, and then, if necessary, an adult hunt. The whitecoat hunt did indeed remain the initial focus, but as the whitecoats began to moult, attention shifted to hoods, both adults and bluebacks. It was only after this phase that the hunters looked to beaters (usually harps), and then only if they had not met their quotas.[46] The few adult harps taken under the quota regime were usually shot incidentally during the whitecoat hunt. These modifications stemmed from the changed emphasis within the industry from fat to furs. Since hood pelts, whether from adults or bluebacks, were often more than twice as valuable as beater pelts, it made economic sense to take as many hoods as possible before taking beaters. Also, the 1977 sealing regulation limiting the catch of adult harp seals to a maximum of five per cent of the total catch increased the relative importance of adult hoods.

Given the modest hood quotas, each vessel only needed to carry two gunners. Sealing regulations introduced in 1967 specified that the gun had to be "a rifle firing only centre fire cartridges, not made with metal cased hard point bullets." The rifle had to possess a muzzle velocity of not less than 1800 feet per second, and a muzzle energy of not less than 1100 foot-pounds. Gunners on Karlsen ships used .303 rifles.

During the motor-vessel era, gunners did not use punts or dories to reach the hoods owing to the earlier start of this phase of the hunt (late March instead of mid-April). Hoods keep towards the seaward edge of the ice field where the stronger and heavier arctic ice is found. Thus the gunners normally went off on foot in search of their prey.[47] Occasionally, when there were only a few isolated hoods near the ships, the gunners would shoot them from the decks.[48] The gunners shot almost exclusively adult male hoods (dogs), aiming at the head so as to avoid damaging the pelt. The 1977 sealing regulations specified that female hoods one year of age or older could account for a maximum of ten per cent of a vessel's total catch of hoods, so bitches were rarely killed. Once again, conservation concerns influenced the hunt.

When the dog was hit, a pair of sealers moved in. One attempted to confuse and distract the bitch, often by waving his tow rope in front of her, thereby giving his partner a chance to drag the blueback away and kill it with his hakapik. This aspect of the hunt was a frequent cause of complaint by animal welfare groups.[49] Bluebacks were usually sculped on the ice and their pelts fixed to a whipline to be towed back to the ship. Sometimes the bluebacks were brought aboard round and sculped on deck so their carcasses could be retained for sale when the ship returned to port. Dogs were more likely to be sculped on the ship, probably because the high value of the pelt necessitated greater care in the operation. Before the dog was winched aboard, it had to be struck with a hakapik to ensure it was dead. It took two men 15 to 20 minutes to sculp a dog hood. The pelts were treated in the same manner as whitecoat pelts. By 1982, hoods comprised approximately ten per cent of the total catch of Newfoundland sealing vessels.[50]

The hood hunt was a lot easier on the sealers. This was partly because they got a chance to rest while the ship moved from the whitecoat ice to the hood ice and because the end of voyage was near. But the main reason was that the hood hunt was not as physically demanding as the whitecoat hunt. There was some risk — every time a man leapt from the ship onto the ice he took a risk, and he faced danger in confronting a hood bitch determined to protect her pup. But generally there was not enough work to occupy all the men. It was not uncommon for a number of them to gather around just to watch the gunners shoot. Those who did work followed no formal division of labour. Instead, the men did whatever jobs they preferred to do, and even changed jobs if they became bored.

The Aerial Spotting Service

The aerial spotting service was revived in 1947 to assist Canadian and Newfoundland sealing ships at the Front and in the Gulf of St. Lawrence. All flights were land-based, the actual departure points being determined by the weather. The gulf planes operated out of the Magdalen Islands and Summerside, Prince Edward Island, and the Front planes out of Gander, Deer Lake, or Goose Bay, Labrador.[51] Until 1955 the airplanes were supplied by Maritime Central Airways of Charlottetown, Prince Edward Island. From that date to 1975 they were provided by Eastern Provincial Airways (EPA purchased MCA in 1963).

In the early years of the postwar spotting service, the most commonly used planes were the Canso flying boat (PBY) and the DC-3. The PBY was more popular because its high wing allowed an unobstructed view of the ice. As catches began to decline, the amount of money paid for the service declined accordingly. In the late 1960s EPA therefore shifted to smaller planes such as the Beech Baron, which cost less to operate. After 1975 EPA withdrew from the spotting service because it possessed no suitable planes. From 1976 until 1982, when it was last employed, the service was provided by smaller companies such as Labrador Airways.

The first flight of each season usually took place around 5 to 7 March.[52] Once the seals were located, daily flights kept track of them until the fleet arrived. Each flight lasted four to eight hours. After the ships had been in a patch for a few days, a new round of flights went out to locate more seals. The spotters radioed the seals' bearings and distance to the captains. Because not all sealing companies were parties to the contract, the spotters frequently used code. If the radios failed, directions were written on a note and the note, attached to a weighted red streamer, was dropped to one of the ships.

The method of payment for the spotting service differed in one key regard from the system used in the 1920s. The owners no longer decided what portion of the catch was attributable to the service. Instead, they paid the contractor a fixed amount per seal based on each ship's total catch. This reflected the complete acceptance of the spotting service as part of the vessel operation. That acceptance came about as a result of several factors, not the least of which was the breakdown of prejudice against the use of airplanes. Even after the Second World War some captains doubted that the seals would be where the spotters said they were.[53] This prejudice was overcome by industry personnel's participation in the service. Most flights carried a pilot and co-pilot and at least one industry representative who was officially designated the seal spotter. Also, on the last flight before the ships left port, a number of the owners themselves went aloft. Another key consideration was the increased ability of the sealing vessels to reach the seals. The spotting service was last used during the 1982 season.

Health and Safety

By the mid-1950s all facets of sealers' safety and shipboard existence had dramatically improved. The first decade after the war had been a period of transition during which some auxiliary schooners and many small motor vessels went down. Even steel ships were not invulnerable. In 1949 two Norwegian steel sealing ships sank in stormy seas, taking 90 men with them.[54] However, such instances were exceptional, whereas the Newfoundland motor vessels were genuinely accident prone. In 1948 alone, the *J.H. Blackmore* succumbed to heavy ice in Bonavista Bay, the *Monica R. Walters* (an auxiliary schooner) and the *Teazer* were both lost in the Gulf of St. Lawrence, and the *Norma L. Conrad* was so severely damaged at the Front that it had to be towed back to port.[55] In 1949 the *Wimoda* was crushed off Belle Isle and its crew rescued by another ship. Two ships were lost in 1951, the *Ranger* in a storm off Baccalieu Island, and the *Lady MacDonald* in heavy ice in the Strait of Belle Isle. Two more casualties were recorded in 1954. The *Newfoundlander*, built by J.H. Blackmore to replace the *J.H. Blackmore*, was crushed in the Strait of Belle Isle; the crew of over 60 men walked across the ice to safety at Eddies Cove. In the gulf the *James Spurrell* struck a ledge and sunk off Codroy. Its small crew of 15 managed to row ashore in dories. The amazing thing about all these accidents is that no lives were lost.

The dreadful performance of the early motor vessels proved to be a temporary phenomenon. Matters improved to the point where it could be said that the safety record of the modern hunt was "excellent."[56] This represented the culmination of several factors, including improved communications and superior ships. Ironically, these gains occurred when compensation benefits were also being extended. Some classes of Newfoundland fishermen came under the purview of workers' compensation legislation in 1948,[57] but sealers were excluded until 1967. Although the Workmen's Compensation (Amendment) Act, 1966-67, did exclude sealers from automatic coverage, upon special application and if accepted by the Workmen's Compensation Board they could receive the same benefits as other classes of fishermen.[58] By the mid-seventies most vessel owners applied for coverage.[59] Coverage for sealers was finally made automatic on 10 October 1980 with the filing of the Workers' Compensation Fishing Regulations, 1980.[60]

The gains in safety were matched by advances in the sealers' shipboard life. Nowhere was this more evident than in their diet. In 1954 the *Arctic Prowler* (Halifax-based but with a Newfoundland crew) included among its provisions roast beef, pork, chicken, smoked and filleted cod, halibut, corned beef, eggs, bacon, dry cereals, tinned and fresh fruit, and tinned milk.[61] During the voyage a typical breakfast consisted of baked beans or bacon and eggs, porridge or dry cereal, and tea or coffee. On Tuesdays, Thursdays, and Saturdays the midday meal was either roast beef or stew — occasionally corned beef and cabbage — with potatoes, turnips, and sometimes corn. Fish was served with potatoes, carrots, and peas on Wednesdays and Fridays. Chicken was the main course on Sundays. Dessert for all meals was custard, pie, and the perennial favourite, duff. For the evening meal, macaroni and cheese, and hash, sausages, and weiners alternated through the week with the exception of Sundays, when meatballs and gravy were served. At night the men could snack on sandwiches they made with leftover meat. In 1953, apparently for the first time ever, men went onto the ice with lunchboxes filled with sandwiches, fresh fruit, and thermoses of tea or coffee.

Sanitation improved, thanks to the continuous availability of fresh water. It appeared on some Halifax vessels as early as the mid-fifties, but seems not to have been standard on Newfoundland ships until a decade later.[62] On the *Lady Johnson*, condensation generated by stove heat was collected in a tank and the water drawn off through a hose. This not only provided a steady supply of drinking water, but also water for washing, a luxury in earlier times. By the 1970s some vessels carried enough water for the entire trip, while others made their own out of condensation from engine heat. Sealing

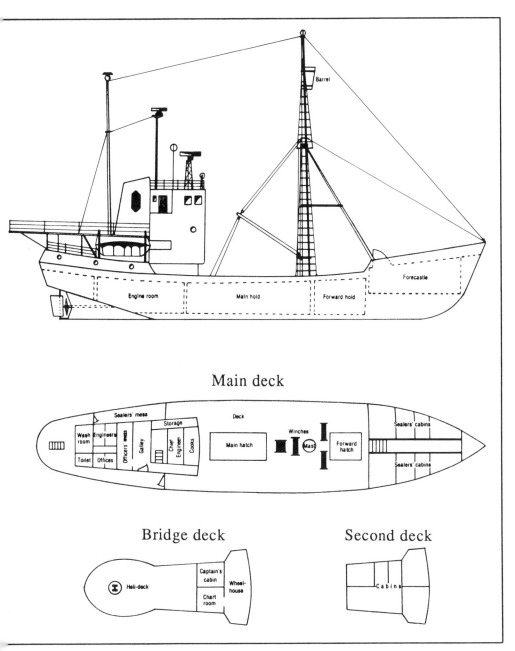

Layout of a sealing ship, 1979.

er Guy David Wright, *Sons and Seals: A Voyage to the Ice* (St. John's: Institute of Social and
ɔnomic Research, Memorial University of Newfoundland, 1984), pp. 30–31.

ships, formerly notorious for their incredible filth, actually had shower facilities.

By 1967 it was possible for a veteran sealer on the *Chesley A. Crosbie* to state that modern ships, compared with even the best steamers, were "like hotels ... with no coal to shovel and better food and living quarters."[63] Wright's account of his voyage aboard the *Hector* (a pseudonym) in 1979 contains the most complete description of living conditions on a modern ship.[64] Unlike the veteran on the *Chesley A. Crosbie*, Wright tended to be less generous in his assessment. The sealers' berths were located in the bow, with four men assigned to each cabin. Each bunk was six feet long and came with a foam mattress. Wright was not impressed by his grimy mattress, but things had progressed from prewar days when the sealer supplied the stuffing for his own mattress, the infamous "donkey's breakfast." The modern sealer was, however, expected to supply his own sleeping bag. The wall next to Wright's bunk was damp from condensation, and the cabin was generally "dark, stuffy, hot and smelled of wet wool, stale sweat and sleeping men."

The modern sealing ship also contained a sealer's mess. No longer did the men have to eat in their berths or make trips to the galley to fetch their food.

> *The sealers' mess was long and narrow, and held four tables that sat six men each. It was separated from the galley by a bulkhead, and plates of food served up in the galley were passed through a small opening to men waiting in line for their meal. After dinner each man had to scrape the remains off his plate into a bin at the rear of the mess, then take the plate back to the galley to be washed.*[65]

The galley itself was quite small, with barely enough room for the cook and cookee to move about. The officers' mess was through a door off the galley.

While giant strides had been made in safety and comfort, sealers still faced danger. Whenever a man jumped off or onto his ship, rolling with the ocean swell, he ran the risk of mistiming his jump and falling between the ship and the ice, where he could be crushed to death. Wright twice saw men miscalculate and land in the water; fortunately both were rescued. When a sealer was crushed between his ship and the ice in 1980, it was remarkable for being the first fatality at the ice "for many years."[66] Men also faced the usual threat of becoming separated from their ship while they were on the ice. However, chances of such accidents were greatly reduced by better communication between ships and up-to-date weather information, and because the men worked closer to their ship than in the steam era.

Much truth lies in the observation that during the modern vessel-based hunt, food and accommodations were superior, but the work was as strenuous as ever.[67]

Processing

In 1949 Karlsen Shipping Company Limited opened a processing plant at New Harbour, Nova Scotia, and in spite of the distance factor, became a competitor for pelts taken by Newfoundland sealers. Job Brothers and Bow-rings closed their plants after they withdrew from the industry in 1952 and 1967 respectively. Their place was taken by Carino Company, which opened a plant at Dildo, Trinity Bay, in 1969. Thereafter, Carino remained the only significant Newfoundland-based processor.

During the 1940s, canned seal meat emerged as a limited commercial product within Newfoundland. Some factories were canning seal meat prior to 1944, but in June of that year the government introduced a licensing and inspection system, which also facilitated accurate record keeping. For the period 1 December 1944–31 May 1945, 34 licences were issued.[68] The 34 canneries processed 32 280 pounds of seal meat, marketed in 48-pound cases, for a total of 672.5 cases. In 1946, 43 licensed canneries provided short-term employment for 106 people. That year's output was 714 cases. The number of factories dropped slightly to 42 in 1947, but production increased to 786 cases. In 1948, output plummeted to 435 cases as the number of licensed canneries fell to 37.

The chief products of the industry remained skins and oil. In 1946 Newfoundland exported 53 132 sealskins worth $403 400 and 180 007 gallons of seal oil valued at $199 300.[69] Skins therefore accounted for 66.9 per cent of seal-related export earnings and oil for 33.1 per cent. This was subject to annual variations: during the 1950s the usual breakdown was 50 per cent oil, 25 per cent leather, and 25 per cent furs.[70] The fur trade now sought not only whitecoats, but also bluebacks, beaters, and both harp and hood adults.

Markets changed during the war years. On the eve of the war, 72 per cent of Newfoundland's sealskin exports went to the United Kingdom, 26 per cent to the United States, and 2 per cent to other countries.[71] By 1946 the United States had taken over as the most important market, accounting for 61 per cent of Newfoundland's sealskin exports. The United Kingdom had slipped to third place with 18 per cent, having been overtaken by Canada at

66 *Opposite top:* Washing pelts, Carino Company plant, Dildo, 1978.
Decks Awash Magazine, Extension Service, Memorial University of Newfoundland, St. John's.

67 *Opposite bottom:* Pelts on a conveyor belt moving toward the fat-removal machine, Carino Company plant, Dildo, 1978.
Decks Awash Magazine, Extension Service, Memorial University of Newfoundland, St. John's.

68 *Above:* Pelts emerging from the fat-removal machine, Carino Company plant, Dildo, 1978.
Decks Awash Magazine, Extension Service, Memorial University of Newfoundland, St. John's.

69 The second fat-removal machine, Carino Company plant, Dildo, 1978.
Decks Awash Magazine, Extension Service, Memorial University of Newfoundland, St. John's.

70 Grading pelts, Carino Company plant, Dildo, 1978.
Decks Awash Magazine, Extension Service, Memorial University of Newfoundland, St. John's.

21 per cent. The markets for seal oil were similarly transformed, thanks to the diversion of all Newfoundland seal oil to Canada to aid the war effort. In 1946 Canada was virtually the only market for seal oil, consuming 98.5 per cent of Newfoundland's exports.

Recent processing methods differ slightly from those of the 1920s. We may take as an example the operation of the Carino Company plant at Dildo in the late 1970s.[72] When the pelts were delivered to the plant, they were submerged in a tank of warm water for thawing and cleaning. They were then placed on a conveyor belt and passed through two rollers that squeezed most of the fat from each pelt. (This was a departure from the earlier practice of separating the fat from the skin by means of a sharp-bladed machine.) A second trip, through another fat-removing machine, squeezed out any remaining fat. Next the pelts were thrown into large, rotating drums and tumbled for about an hour in special hardwood sawdust that absorbed dirt and oil. After this final cleaning, the pelts (skins with fur attached) were ready for grading and, finally, storage in barrels. As always, generous amounts of salt were spread over each skin.

Both the Carino plant at Dildo and the Karlsen plant at New Harbour possessed facilities for rendering seal fat into oil. As before, the fat from several hundred skins was placed in a digester and cooked by steam injected under pressure. Seal fat is very low in fibre, consisting of 80 per cent oil.[73] After the fat had cooked for an hour the oil was drawn off, ready for export. The old method of bleaching the oil in sunlight was not part of the modern operation. Carino-manufactured seal oil was usually shipped to Europe, while Karlsen sent its oil to Canadian and European buyers. The oil was subsequently used in making chocolates, margarine, machinery lubricants, cosmetics, and edible oil products such as non-dairy whipping cream. The value added to seal oil after it left Newfoundland far exceeded its value at point of export. Seal oil contributed an estimated $610 400 to Atlantic Canada's economy in 1980 but achieved a final market value of over $4 million.[74] When one considers that this had been going on since the origin of the commercial seal hunt in the 18th century, the economic implications for Newfoundland are numbing, although hardly unique in Canadian economic history.

Sealskins became important to the fur trade beginning in the 1920s, and by the 1950s the ratio of the value of fur products to leather products was 50/50. That ratio changed dramatically in the 1970s when quotas were imposed. Because the number of seals available to each ship was limited, sound economics dictated that sealers should first take the pelts that commanded the best price: those on which the fur was still fast to the skin. Previously sealers had not been so discriminating, knowing that raggedy jackets (moulting whitecoats), useless for their fur, could be used for leather. Since raggedy jackets were now less valuable, they were taken only if better pelts could not be found; given the quotas, this did not often happen.[75] Thus, instead of a 50/50 ratio of fur to leather, the ratio in the last decade of the hunt was 95/5.

By the mid-seventies, significant change had occurred in the relative importance of seal products.[76] For landsmen, pelt sales contributed 77 per cent of gross revenues at the primary, sealer-to-processor level compared with 14 per cent and 9 per cent for meat and oil respectively. The industry's overwhelming dependence on pelt sales was a contributing factor in the success of the protest movement, which focussed on fur products. Equally noteworthy was the position of meat sales (although beyond the primary sales level, oil was still more important). The three plants in Newfoundland processing seal meat in 1976 employed 15 to 18 people over a maximum of 21 weeks.[77] Total production came to 128 000 pounds, marketed in 7-ounce and 14-ounce tins, nearly four times 1945 production. Sales of fresh and frozen seal meat were actually more significant by this date. Only 6 per cent of the total catch in 1976 was canned, whereas sales of flippers and carcasses accounted for 22 per cent of total seals killed and generated revenues of $490 000. The market for seal meat remained entirely local.

What happened to the pelts after they left Newfoundland? The answer depends on the date and reflects the shift in control of the industry from British to Norwegian interests. In 1936 Bowrings established a working relationship with W. Duckworth and Sons, Limited, which produced tanned seal leather for the British shoe trade.[78] Agents based in London acted as middlemen between Duckworth and the trade. In 1947 Bowrings created its own leather and fur department to handle all sales of sealskin products, retaining Duckworth as its tanner. Duckworth moved away from shoe leather in 1948 and began producing for the fancy leather trade. This product was exported to manufacturers around the world, notably in the United States, West Germany, Holland, France, Sweden, Portugal, Canada, Mexico, and Australia. Although Duckworth's tannery did dress and dye some seal furs, that part of the business was generally contracted out and the furs marketed through London fur brokers. There was very little waste. Fur shaved from skins earmarked for the leather trade was sold to the garment industry to be blended with wool for coats. Duckworth became a wholly owned Bowring subsidiary in 1952.

Carino Company was a subsidiary of G.C. Rieber and Company of Bergen, Norway, and it was there that pelts were shipped after 1969. At Bergen the pelts were subjected to further cleaning in drums, followed by drying, stretching, tanning, and yet more drumming.[79] The actual tanning, by which the hides were converted into leather, took only 24 hours; most of the seven-week period was taken up by drumming and drying. From Bergen the pelts were exported to any of 30 countries, but primarily to Denmark, West Germany, and other west European countries (until the EEC ban). Fur manufacturers in those countries added more value by producing finished fur coats. Indeed, for the 1980 season, revenues generated in Europe are estimated to have exceeded by three to six times the value of the pelts at point of export in Newfoundland.[80] Carino Company did install a sealskin dressing facility at the Dildo plant in 1981, but it was a classic case of too little, too late.

Pay

During the Second World War, the seed was sown for an entirely new method of paying sealers. When the Karlsens came to Halifax they introduced the Norwegian system of basing shares on the value of the catch: the sealers shared 26 per cent of the graded value of the pelts (the grader being a company employee) less processing costs before export.[81] Newfoundland sealing companies were slow to adopt the new system because it resulted in higher pay to the sealers. Also, the Newfoundland system had the advantage of keeping the sealers ignorant of the true value of the catch. Newfoundland sealers serving on Nova Scotian ships earned twice as much as those who sailed with the Newfoundland fleet.[82] Other factors, such as superior ships, undoubtedly contributed to this gap, but the pay structure was an important consideration. It has not been possible to identify when Newfoundland firms finally adopted the Nova Scotian/Norwegian system, but they appear to have done so by the early 1960s. By that time, 80 per cent of Newfoundland vessel-based sealers were sailing on Nova Scotian ships.

In the immediate postwar years, share size continued its upward climb as a result of the high price of seal oil and, more importantly, the small size of the labour force. When a record average share of $186 was achieved in 1947, the three steamers and 15 motor vessels of the Newfoundland fleet carried only 650 sealers.[83] In 1953, shares ranged from $45 to $175, attesting to the venture's continued riskiness.[84] Four years later the comparable figures were $61 and $197.[85]

By the 1950s, share statistics had become more misleading than ever because of the increasing revenue generated by flipper sales. Each sealer was allowed one barrel (which he supplied himself) in which he could store 12 (and occasionally 13) dozen flippers.[86] In 1953, sealers on the first ship back from the hunt commanded $12 per dozen flippers, with the price falling to $4 per dozen by the time all the other ships had arrived.[87] If each man filled his barrel, he would have augmented his share by anywhere from $48 to $144, for maximum average earnings of $319 (excluding deductions for crop and transportation expenses). By 1957, sealers were earning up to $200 from flipper sales alone.[88]

In 1963 the Newfoundland Fisheries Commission estimated average annual earnings from the seal hunt at $400 per man.[89] The next year, the men of the *Algerine* set a new record, making $750 each on the season, which presumably included revenues from flipper sales.[90] This was surpassed in 1967 when sealers on the *Chesley A. Crosbie* received approximately $800 each.[91] In 1970, sealers on the *Chesley A. Crosbie* and the *Lady Johnson*, the only Newfoundland-owned ships, realized approximately $500 apiece.[92] The most striking gains occurred in the 1970s as crew size shrank because of the quota system. The average gross income of the 189 large-vessel-based hunters in 1976 was $1877 for 29 days of work.[93] This compares with $1256 for longliner sealers and $232 for all other landsmen.

The 1976 survey of the east coast sealing industry contains comprehensive statistics on returns to vessel owners.[94] Based on submissions for five of the six large vessels operating that year, the average income per ship from sales of pelts and oil (less processing charges) was $137 463. Expenditures averaged $48 379, leaving an average net cash return of $89 084 per vessel. Expenditures included $15 991 for maintenance and repairs, $26 023 for other operating expenses (fuel, oil, grease, provisions, wages, wharfage, and miscellaneous), and $6365 for fixed charges. Fixed charges included crew insurance and payments to workers' compensation, but the major charge was $4761 for marine insurance, or 10 per cent of all expenses. Of the net cash return of $89 084, $50 841 (57 per cent) was deducted for crew shares, leaving a net cash vessel share of $38 243. Allowing for depreciation of $2924, the average net earnings per vessel were $35 319.

Historically, the highest retuns to labour were achieved in the final years of the hunt. A man's share on Wright's vessel in 1979 amounted to $2197.[95] Sales of flippers and carcasses brought the sealers an additional $747 each, for total individual earnings of $2944. Such a figure would have been unheard of a decade earlier. Of course, there were still some deductions. The men were now charged for the food they consumed on the voyage and for any supplies they might have acquired. Although the word "crop" was no longer used, the practice was still observed in slightly modified form. From a storeroom on the vessel sealers could get anything from soft drinks and cigarettes to gloves, knives, and sharpening steels. The cost of any purchases was deducted from a man's share at the end of the voyage. As well, a sealer had to pay his transportation costs to and from St. John's.

The actual crew settlement was determined in a fairly straightforward manner.[96] Certain costs were first deducted from the ship's gross income: insurance costs, workers' compensation contributions, cost of ammunition the gunners actually used, cook's wage, cook's share (1$\frac{1}{4}$ shares), cost of the spotting plane, and the cost of processing the pelts (skins and fat). The net profit was divided into shares, each sealer receiving one full share; the combined shares of all sealers totalled 26 per cent of the net profit. Before the actual cash vessel share could be determined, several other deductions had to be made: the chief engineer received two shares plus a wage; second engineer, 1$\frac{1}{2}$ shares plus a wage; first mate, two shares plus a wage; second mate, 1$\frac{1}{4}$shares; oiler, 1$\frac{1}{4}$shares; cook's helper, 1$\frac{1}{4}$shares; gunners, 1$\frac{1}{4}$shares. In addition, the captain was entitled to one share plus a percentage of gross earnings. After the payment of shares and wages, the result was roughly a 50/50 split between the owner and crew (sealers and non-sealers).

Conclusion

The seal hunt played a vital role in Newfoundland's economy and culture. To the fishermen of the pre-confederation era, sealing was virtually the only way they could obtain cash because all other branches of the fishery functioned on credit. For the country as a whole, sealing was, in the 19th century, second only to the cod fishery in economic importance. The sealing industry's role had begun to diminish by mid-century, but even then it contributed 25 per cent of the value of all exports and employed nearly 15 000 men. The decline, subtle enough at first, was hastened by the advent of steamships and more sophisticated processing equipment. This intensified pressure on the seal resource and promoted the industry's concentration in St. John's. This latter development helped to undermine Newfoundland's outport economy.

The decline continued early in the 20th century, aided by government and entrepreneurial inaction; imaginative solutions such as an expanded auxiliary-schooner operation were never given a proper chance, likely because they would have threatened the St. John's monopoly. Attempts to create a manufacturing sector of the economy and to attract foreign capital to develop forest and mineral resources entailed neglect of the fishery, including sealing.

Meanwhile, the seal herds were rejuvenated because of reduced hunting. With the Second World War putting nearly a complete stop to the hunt, Newfoundland was presented with a golden development opportunity. Instead, that opportunity was seized by Norwegian and Canadian interests while the provincial government exhausted its surplus funds accumulated during the war years in yet another futile attempt to expand the manufacturing sector. Any money for the fishery was funnelled into the new fresh- and frozen-fish sector, with the blessings of the federal and provincial governments, and again traditional branches of the fishery were neglected. Some consolation was found in Nova Scotia–based sealing companies' need for Newfoundland labour.

The bleak picture finally began to improve after 1977 when Canada adopted a 200-mile fishing zone and acquired complete jurisdiction over the resource. Through the reduction of their allotment, the Norwegians were ultimately squeezed out, leaving a void that Newfoundland companies started to fill. But at the very moment when Newfoundland was re-establishing its historic role, the industry was struck a crippling, perhaps life-ending blow in the form of the EEC boycott of harp and hood seal-pup products. In 1984, animal welfare groups organized consumer boycotts of Canadian fish products in Great Britain and the United States in an attempt to put a complete stop to the commercial seal hunt. In the face of this threat, no ships went to the ice in 1984 and the landsman hunt was drastically curtailed. The industry's fortunes now depend on the discovery of new markets or the repeal of the EEC boycott.

The attack on the seal hunt has generated vigorous defences from Newfoundlanders, who base their arguments primarily on economic and cultural grounds.[1] Although they know that sealing is no longer of overall importance to the provincial economy, they stress that it is still important to individual communities along the northeast coast. Because of the seasonal nature of the fishery and the absence of alternative employment, sealing is in these places the only source of income, other than welfare or unemployment insurance payments, during the winter and spring. For some fishermen, sealing can contribute up to one-third of their annual income. Money earned from sealing can be used to reduce or pay off debts, or to purchase new equipment for other branches of the fishery. In a province as economically depressed as Newfoundland, they argue, an industry that provides employment for four thousand people is an industry worth keeping.[2]

The morality of killing seals is a question that did not occur to the native peoples and European settlers because if they did not kill seals they themselves might have perished. The seal was not a special case: wild animals, birds, and fish were also taken. Over time this pattern became ingrained in the life of outport Newfoundland. Killing animals "was, is, and must remain" the cornerstone of the outport economy.[3]

One result of the international attack on sealing has been a tendency within Newfoundland to romanticize the industry. In truth, the industry promoted underdevelopment of the economy, especially after it became concentrated in St. John's. Only the most preliminary stages of processing seal oil and skins ever took place in Newfoundland. Most processing occurred in Europe, and Europe was where the bulk of the profits accrued. A recent study concluded that while sealing contributed $12 117 000 to the gross regional product of Atlantic Canada in 1980, expansion of the industry, especially the processing sector, could have meant an additional $25 million.[4] This weakness in the processing sector had existed since the origin of commercial sealing in the 18th century. The sealing companies also grossly exploited their labour force. Pay was low, working conditions were abysmal, and compensation was minimal. Most of the advances in these areas, pay in particular, took place only after the Second World War.

Try as they might, Newfoundlanders cannot convince the outside world that the seal hunt is a legitimate enterprise. Why is the anti-sealing campaign so successful? Certainly the distance factor has contributed to a one-sided debate.[5] The scene of the hunt is remote from the protest centres, and as a consequence the protestors' charges are largely uncontested in those centres. The Newfoundland government's 1978 pro-sealing campaign was the only concerted effort to defend the hunt before an international audience. It has been argued that the federal government should have mounted a defence in the 1960s when the protest first began.[6]

Sealer/journalist Pol Chantraine attributes the protestors' success to the subconscious appeal of harp seal pups.[7] His interpretation borrows from the work of zoologist and ethologist Konrad Lorenz, who theorized that certain physiological features of humans and animals promote specific behavioural responses. Lorenz called these features innate releasing mechanisms. Colour, for example, is a strong releasing mechanism — a male robin will attack a cluster of red feathers, treating it as a rival. Chantraine saw that a whitecoat shares many of the same characteristics that adults respond to in a child: proportionately large head, large low-lying eyes, and awkward movements. He concluded that the physical appearance of the whitecoat subconsciously triggers protective behaviour among humans. A fish, by contrast, possesses few features to which a human will respond positively. Thus seal-hunt protestors ignored the depletion of commercial fish stocks in the waters of Atlantic Canada, even though their plight was comparable to that of the harp seals.

The controversy over the seal hunt is, more than anything else, a clash of cultures. Broadly speaking, the protestors represent modern, urban culture and the sealers the traditional culture of the past. A sentimental view of the natural world is central to the modern sensibility. Before the industrial revolution, most people regarded animals as necessary for their survival, a belief reinforced by the Christian Bible, which taught that God gave man dominion over beasts.[8] This world-view began to break down by the late 18th century as scientific discoveries challenged the notion that the earth and all things in it, including animals, had been created exclusively for the benefit of humans. For example, it came as a revelation when anatomists discovered structural similarities between human and animal bodies. Some were even bold enough to suggest that humans were descended from animals. The old standards started to crumble.

The scientific attack on the anthropocentric tradition coincided with a shift in moral attitudes towards animals. Liberal theologians fostered a wave of human compassion in the 18th century when they claimed that salvation could be achieved through helping one's fellow man.[9] This philosophy, known as humanitarianism, had its most obvious manifestation in the anti-slavery movement. Inevitably, compassion spilled over to animals. In 1780 utilitarian philosopher Jeremy Bentham specifically likened the position of animals to that of slaves, arguing that it did not matter whether animals possessed souls or were capable of reason. Simply that they could experience suffering made them worthy of moral consideration.[10] It was no accident that many early animal protectionists were also active abolitionists.

The industrial revolution reinforced the new scientific and moral approaches to humanity's relationship with the natural world. As an urban, industrial society replaced a rural, agrarian one, the functional role of animals was reduced, making it easier to view them sentimentally. The change ran even deeper than that. Rural landscape and rural culture also became romanticized, witness the poetry of Wordsworth, the philosophy of Thoreau, and the art of Constable, among others. The gap between humans and animals continued to close and was finally obliterated in 1859 with the publication of Charles Darwin's *Origin of Species*, which shattered the myth of humankind's exclusivity. Darwin's theories made it all the more imperative, at least among the educated classes, that humans be kind to animals: cruelty to animals was evidence of our own animality.[11]

After Darwin, compassion for animals began to merge with a more sophisticated view of humankind's relationship with nature.[12] Instead of having dominion over nature, humans were now perceived to be part of it. This promoted the concept of the balance of nature, which held that all species were interdependent. Humans had an obligation to preserve other species because the loss of one would ultimately affect all. This formed the scientific and philosophical bases of the current ecological movement, which originated in the 1960s when the DDT issue drew attention to the passage of pesticides through the food chain, thus emphasizing once again the interdependence of all nature.

The seal-hunt protest movement embodies both components of the modern attitude towards nature. The modern sensibility incorporates concern for cruelty to animals with a commitment to the preservation of species. The first is the focal point of IFAW and the second of Greenpeace. Significantly, the protest groups have their greatest appeal in western industrialized nations: Great Britain, France, West Germany, and the United States. Even within Canada, opposition to the hunt is strongest in Ontario, the most industrialized province.[13] Unlike the protestors, the sealers live in areas that have not yet completely succumbed to industrialism. Their view of nature is essentially that of a pre-industrial people, and the pre-industrial economy "almost always concerns primary production and resource exploitation."[14]

It is not enough to dismiss the protestors as victims of "mawkishness."[15] The protest movement is symptomatic of changing attitudes towards our relationship with the natural world, typified by the emergence of the animals' rights movement. Advocacy of rights for animals is hardly new, but recently it has taken on a higher public profile.[16] Animals' rights activist Richard Ryder has coined the word "speciesism" to describe "mankind's arrogant prejudice against other species."[17] Ryder places speciesism in the context of racism and sexism:

> *Today, animals are by far the most oppressed section of the community: their exploitation is as great an evil as were black slavery, child labour and the degradation of women at the beginning of the last century. It is the great moral blind spot of our age.[18]*

What makes animals worthy of such moral consideration? Modern proponents of animals' rights echo Jeremy Bentham when they argue that the ability to experience pain or pleasure puts animals on the same moral plane as humans.[19] Philosopher Peter Singer also uses analogies of racism and sexism:

> *The racist violates the principle of equality by giving greater weight to the interests of members of his own race when there is a clash between their interests and the interests of those of another race. The sexist violates the principle of equality by favoring the interests of his own sex. Similarly the speciesist allows the inter-*

ests of his own species to override the greater interests of members of other species.[20]

For the animal liberationist, killing animals to make fur coats is not justifiable because it violates the interests of those animals. But that is only one area of concern and not the most important. The animal liberationist also opposes factory farming and using animals in research, which are deemed to cause the majority of animal suffering.[21] In Great Britain alone, some five million experiments are performed on animals each year, many of them, according to Singer, "without the remotest prospect of significant benefits for humans or any other animals." Even more objectionable to the animal liberationist is using animals to test for the toxicity of cosmetics, shampoos, floor wax, and other domestic products. It does not matter if such products are not lethal in small doses, for researchers generally seek to determine "Lethal Dosage 50," the dosage level necessary to kill 50 per cent of the animals being tested. Again according to Singer, up to one million animals a year in Great Britain alone die as a result of cosmetic research.

It is important to distinguish here between animal liberation and animal welfare. Animal welfare groups have traditionally been largely silent on the issues of factory farming and research involving animals. Singer goes so far as to charge that animal welfare societies contribute to animal suffering by turning strays over to laboratories where they are used in experiments.[22] Also, true animal liberationists are vegetarians and eschew the wearing of fur and leather, for if they did otherwise they would be morally inconsistent. It is highly unlikely that the people who contributed over five million dollars to IFAW in 1983 observe such exacting standards.[23] Most of them probably wear leather, use products that have been tested on animals, and, like Brian Davies, eat meat. A prime example of such double standards occurred in March 1984, when Brigitte Bardot tried to donate an ivory bracelet to the Quebec SPCA, only to have it seized by Canada Customs because the ivory came from an African elephant, itself an endangered species. Historian Keith Thomas suggests it is no coincidence that as barriers between humans and animals broke down, the "concealment of slaughter-houses from the public eye had become a necessary device to avoid too blatant a clash between material facts and private sensibilities."[24] That conflict, he argues, has never been resolved and is "one of the contradictions upon which modern civilization may be said to rest."[25] But unlike the bulk of humanity, animal liberationists have made their decision.

Both Greenpeace and IFAW have moved away from their respective ecological and animal-welfare perspectives and into the animals' rights camp. This was evident during the opening hearings of the Royal Commission on Seals and the Sealing Industry in Canada when Richard Ryder appeared as a consultant to IFAW. This new coalition is now taking aim at the Canadian fur industry. While they may even be successful in this next campaign, it is difficult to imagine that possible campaigns against meat eating and using animals in research will generate the same widespread response that characterized the seal-hunt protest. Meat eating cuts across every socio-economic boundary and is such an integral part of national economies that its elimination would involve massive cultural and economic restructuring. And a 1974 survey conducted by the RSPCA revealed that 80 per cent of British voters accepted the use of animals in medical research.[26] This points out the convenience of seal-hunt protest: it has given people in the industrial world an outlet for guilt feelings about the way animals are treated without forcing them to make serious adjustments to their lives. Any adjusting has been left to the sealers, like the landsmen who have lost one-third of their annual income and can no longer meet the payments on their longliners. Their whole livelihood is now in jeopardy because without longliners they cannot take part in the summer cod fishery.

Animal liberation is gaining momentum. Current research involving the transfer of human genetic material to pigs, sheep, and other animals has prompted a vigorous ethical debate that may open the whole issue of laboratory research on animals. In November 1984 the radical Animal Liberation Front focussed attention on that issue when they put rat poison in Mars candy bars in five British cities to protest the use of monkeys in Mars-funded research into the connection between candy and tooth decay. On 1 January 1985, members of the Animal Liberation Front broke into a laboratory at the University of Western Ontario and, depending on one's perspective, stole or liberated a rhesus monkey and three cats being used in medical research. The Baby Fae case, in which a baboon heart was transplanted into an infant (who later died) has likewise generated controversy. This was a clear case of speciesism, in which the interests of the human were placed above those of the donor animal.

No one knows whether the general public will some day embrace the strict moral discipline of animal liberation. But public response will prove whether opposition to the seal hunt was an important milestone in the fight against speciesism, or just a massive exercise in moral inconsistency.

Afterword
Update to January 1988

Since mid-1985, when the preceding chapters were written, there have been several new developments in the seal-hunt story. These include the submission of the report of the Royal Commission on Seals and the Sealing Industry in Canada, a revival of the landsman hunt, and the banning of the vessel-based hunt.

The Royal Commission on Seals and the Sealing Industry in Canada conducted hearings from January to May 1985. After a lengthy waiting period highlighted by numerous press leaks and an injunction against the Canadian Broadcasting Corporation, the report was finally tabled in the House of Commons on 17 December 1986. Its main recommendations were an official end to hunting seal pups; a controlled hunt of adult seals with a quota of 20 000 to 60 000 pelts depending on market conditions; payment of $50 million to Newfoundland, Quebec, and Maritime sealers as compensation for lost income and a further $20 million to Inuit sealers for the same purpose; and establishment of a $50-million fund for economic development in sealing communities. Strangely, while the commission recommended an official end to the seal-pup hunt, it also found that the EEC ban was unfair because the seals were killed humanely and were not endangered. The Canadian government responded by announcing that it would not officially end the hunt, but neither would it encourage its resumption. It would not grant compensation for income loss, but instead would spend just over $5 million in an attempt to find new markets for seal products.

The royal commission report was hailed by both the Canadian Sealers' Association (CSA) and protest groups. Since the elimination of the pup hunt would affect only vessel-based sealing, landsmen, as represented by the CSA, were delighted. And all sealers could take some solace from the commission's finding that the hunt was indeed humane. Likewise, the protest groups viewed the recommendation to end the pup hunt as vindication of their efforts to stop it.

The landsman hunt reached rock bottom in 1985 and 1986, when only some 18 000 seals were taken. Nevertheless, 1986 marked the beginning of a renewed effort by landsmen, backed by government, to revive the industry. In April a new seal-pelt-processing plant opened at Fleur de Lys on the Baie Verte Peninsula, in the heart of landsman country. The Newfoundland government gave the plant's operators, the Northeast Coast Sealers' Co-operative, a $25 000 grant and a $200 000 loan guarantee, while 350 landsmen put up $300 each to become co-op members. The federal government subsequently chipped in with a $49 200 grant. The CSA revealed that the opening of the plant was part of a strategy to rebuild the industry over eight to ten years. That strategy also included market studies in Taiwan, Hong Kong, and Japan. The Fleur de Lys plant processed approximately 11 000 adult seal pelts in 1986, most of which found their way to central Canada for use in manufacturing boots and handicraft items.

Early in 1987 the Newfoundland government increased its financial support to the industry. It gave the Northeast Coast Sealers' Co-operative $1 million ($700 000 of it in an interest-free loan) towards the costs of purchasing and processing pelts during the year. It also issued grants totalling $70 000 to the CSA for studies on the potential use of seal pelts in Newfoundland crafts and on markets for canned and frozen seal meat. Buoyed by the infusion of government money, Carino Company announced in March that it was re-opening its Dildo plant after a three-year shutdown.

Up to this point it had been assumed that because of the EEC ban on seal-pup products, the vessel-based hunt was dead. It had been a truism within the industry that it was uneconomical for large vessels to concentrate on anything other than whitecoats. Thus it came as a shock when Karlsen Shipping Company and Puddister Trading Company announced their intention to send ships to the Front in 1987 to hunt adult seals. Both the CSA and the Newfoundland government, which had cast its lot with the landsmen, immediately appealed to the federal minister of Fisheries and Oceans, Tom Siddon, to cancel the licences issued to the two firms. They feared that a revival of the vessel-based hunt would arouse the protest groups and thereby jeopardize the progress being made in the landsman hunt. Siddon replied that it was too late in the season to change the government's sealing policies and set a quota of 57 000 seals for the two ships.

The resumption of the vessel-based hunt caught the protest groups completely off guard. Since the cessation of vessel-based sealing in 1984, the protestors had turned their energies elsewhere. Greenpeace was thrust into the international spotlight in July 1985 when French secret-service agents sank the *Rainbow Warrior* as it lay at anchor in Auckland, New Zealand, prior to a voyage to protest French nuclear testing in the South Pacific. Then in November 1986 Paul Watson's Sea Shepherd Conservation Society grabbed headlines when its members sank two Icelandic whaling ships in Reykjavik harbour and damaged an adjacent whaling factory to protest alleged illegal whaling. When it became known that the vessel-based seal hunt was resuming off Newfoundland, Greenpeace was unable to do anything because all its available vessels were still in the Pacific Ocean and could not reach Newfoundland in time to protest the hunt. IFAW and the Sea Shepherd Conservation Society were also unable to respond, but all three groups promised that they would be ready should there be a vessel-based hunt in 1988. The only protesters to make it to the region in 1987 were from the recently created International Wildlife Coalition, led by former IFAW member Dan Morast. They abandoned plans to photograph the hunt when they failed to charter a plane to fly them to the Front.

The 1987 vessel-based hunt was like none other in history except perhaps the shallop-based hunt of the 1790s. No whitecoats were taken; instead, the 20 sealers on the *Chester* and the 18 on the *Terra Nova* used rifles to kill adult seals. And in a radical departure from tradition, sealers on both vessels received salaries instead of shares. The experiment turned out to be a disaster owing to extremely heavy ice at the Front. The two ships between them managed to bring in only some 3100 pelts.

The reappearance of the vessel-based hunt and the subsequent arousal of the protest groups raised once again the spectre of a boycott of Canadian fish products. Accordingly, in December 1987 the federal government took its royal commission's advice and banned sealing by large offshore vessels. This officially ended the hunt at the Front and put the industry completely in the hands of the landsmen. The industry had come full circle.

The 1987 seal hunt had an entirely different outcome for Newfoundland landsmen. The 6000 individual landsmen and 75 longliner crews who received licences took approximately 39 000 pelts, the highest number since 1982 (51 703). It is too early to tell whether the 1987 catch reflects improving market conditions or whether it is the result of artificial demand created by government assistance to the industry. Since the EEC ban is unlikely to be repealed, the industry's long-term prospects depend on the discovery of new markets.

Endnotes

Introduction

1 David Ernest Sergeant, "History and Present Status of Populations of Harp and Hooded Seals," *Biological Conservation*, Vol. 10, No. 2 (1976), pp. 95–101; Chesley W. Sanger, "Technological and Spatial Adaptation in the Newfoundland Seal Fishery During the Nineteenth Century," MA thesis, Memorial Univ. of Newfoundland, St. John's, 1973 (hereafter cited as "Newfoundland Seal Fishery"), pp. 29–86.

2 Captain Morrissey Johnson, "Thirty Years at the Seal Hunt: Experiences of Morrissey Johnson," *The Livyere*, Vol. 1, Nos. 3–4 (Winter/Spring 1982) (hereafter cited as "Thirty Years at the Seal Hunt"), p. 42.

3 Major William Howe Greene, *The Wooden Walls Among the Ice Floes: Telling the Romance of the Newfoundland Seal Fishery* (London: Hutchinson, 1933) (hereafter cited as *Wooden Walls Among the Ice Floes*), pp. 79–80.

4 Ibid., p. 11.

5 Abraham Rees, *The Cyclopaedia; or, Universal Dictionary of Arts, Sciences and Literature* (London: Longman, Hurst, Rees, Orme and Brown, 1819), Vol. 27, s.v. "Phoca."

6 Patrick Thomas McGrath, *Newfoundland in 1911....* (London: Whitehead, Morris, 1911), p. 140.

7 Farley Mowat, *Wake of the Great Sealers* (Toronto: McClelland and Stewart, 1973) (hereafter cited as *Wake of the Great Sealers*), p. 32.

Origins and Early Development

1 Unless otherwise noted, all material in this section is based on James A. Tuck, *Newfoundland and Labrador Prehistory* (Toronto: Van Nostrand Reinhold, 1976). I am also indebted to Charles Lindsay, Historic Resources Research, Atlantic Regional Office, Canadian Parks Service, Environment Canada, Halifax, for his comments on the original draft of this section.

2 William Gilbert Gosling, *Labrador: Its Discovery, Exploration, and Development* (London: Alston Rivers, 1910) (hereafter cited as *Labrador*), p. 17.

3   Lewis Amadeus Anspach, *A History of the Island of Newfoundland....* (London: T. and J. Allman, 1819) (hereafter cited as *A History of the Island of Newfoundland*), p. 417.

4   Selma Barkham, "The Basques: Filling a Gap in our History between Jacques Cartier and Champlain," *Canadian Geographical Journal*, Vol. 96, No. 1 (Feb./Mar. 1978), p. 8.

5   *New American World: A Documentary History of North America to 1612*, ed. David B. Quinn, Alison M. Quinn, and Susan Hiller (New York: Arno Press, 1979), Vol. 4, pp. 56–63.

6   Edward Thomas Davis Chambers, *The Fisheries of the Province of Quebec* (Quebec: Dept. of Colonization, Mines and Fisheries, 1912), Pt. 1, p. 59.

7   H.M. Mosdell, ed., *Chafe's Sealing Book; A History of the Newfoundland Sealfishery from the Earliest Available Records down to and including the Voyage of 1923. Statistics Prepared by L.G. Chafe* (St. John's: Trade Printers and Publishers, 1923) (hereafter cited as *Chafe's Sealing Book*), p. 15.

8   C. Grant Head, *Eighteenth Century Newfoundland: A Geographer's Perspective* (Toronto: McClelland and Stewart, 1976) (hereafter cited as *Eighteenth Century Newfoundland*), p. 77.

9   William Gilbert Gosling, *Labrador*, p. 203.

10   C. Grant Head, *Eighteenth Century Newfoundland*, p. 77.

11   Averil M. Lysaght, ed., *Joseph Banks in Newfoundland and Labrador, 1766: His Diary, Manuscripts and Collections* (Los Angeles: Univ. of California Press, 1971), p. 104.

12   Daniel Woodley Prowse, *A History of Newfoundland from the English, Colonial, and Foreign Records* (St. John's: Dicks, 1971)(hereafter cited as *A History of Newfoundland*), pp. 419–20.

13   *Dictionary of Newfoundland English*, ed. G.M. Story, W.J. Kirwin, and H.D.A. Widdowson (hereafter cited as DNE), s.v. "shallop."

14   Edward Chappell, *Voyage of His Majesty's Ship Rosamond to Newfoundland and the Southern Coast of Labrador....* (London: printed for J. Mawman, 1818), pp. 197–99.

15   Lewis Amadeus Anspach, *A History of the Island of Newfoundland*, pp. 414–15; Sir Richard Henry Bonnycastle, *Newfoundland in 1842: A Sequel to "The Canadas in 1841"* (London: Henry Colburn, 1842) (hereafter cited as *Newfoundland in 1842*), Vol. 2, p. 131.

16   Joseph Hatton and Moses Harvey, *Newfoundland; Its History, Its Present Condition and Its Prospects in the Future....* (Boston: Doyle and Whittle, 1883) (hereafter cited as *Newfoundland*), p. 248.

17   H.M. Mosedell, ed., *Chafe's Sealing Book*, p. 20; DNE, s.v. "shallop."

The Seal Hunt
1793–1861

1   Shannon Ryan, "The Newfoundland Cod Fishery in the Nineteenth Century," MA thesis, Memorial Univ. of Newfoundland, St. John's, 1971 (hereafter cited as "The Newfoundland Cod Fishery"), p. 15.

2   C. Grant Head, *Eighteenth Century Newfoundland*, p. 225.

3   Shannon Ryan, "The Seal and Labrador Cod Fisheries," manuscript on file, No. ARO-0014, Atlantic Regional Office, Candian Parks Service, Environment Canada, Halifax, n.d. (hereafter cited as "The Seal and Labrador Cod Fisheries"), p. 12. A similar version of this manuscript, "The Seal and Labrador Cod Fisheries of Newfoundland," was

published as Vol. 26 of *Canada's Visual History* (Ottawa: National Museum of Man/National Film Board, n.d.).

4 Lewis Amadeus Anspach, *A History of the Island of Newfoundland*, p. 443.
5 By the Treaty of Utrecht (1713) the Newfoundland coast between Cape Bonavista and Pointe Riche had been reserved for French fishermen although they were not allowed to settle there. The Treaty of Versailles (1783) established a new French shore between Cape St. John and Cape Ray as a result of the northward spread of British settlement into Bonavista and Notre Dame Bays during the 18th century.
6 Shannon Ryan, "The Seal and Labrador Cod Fisheries," p. 12.
7 Chesley W. Sanger, "The 19th Century Seal Fishery and the Influence of Scottish Whalemen," *Polar Record*, Vol. 20, No. 126 (Sept. 1980) (hereafter cited as "Influence of Scottish Whalemen"), p. 235; Chesley W. Sanger, "The Evolution of Sealing and the Spread of Permanent Settlement in Northeastern Newfoundland," in John J. Mannion, ed., *The Peopling of Newfoundland: Essays in Historical Geography* (St. John's: Institute of Social and Economic Research, Memorial Univ. of Newfoundland, 1977) (hereafter cited as "The Evolution of Sealing"), p. 150.
8 Chesley W. Sanger, "Influence of Scottish Whalemen," p. 235.
9 Ibid., "The Evolution of Sealing," pp. 136–41.
10 H.M. Mosdell, ed., *Chafe's Sealing Book*, p. 16.
11 Shannon Ryan, "The Seal and Labrador Cod Fisheries," p. 12.
12 *The Newfoundlander* (St. John's), 11 Mar. 1861, p. 3; 14 Mar. 1861, p. 3; 8 April 1861, p. 2.
13 Newfoundland. Colonial Secretary's Office, *Census ... 1845, ... 1857, ... 1869* (St. John's: 1845, 1857, 1870).
14 Chesley W. Sanger, "The Evolution of Sealing," pp. 143–44.
15 Captain Robert Abram Bartlett, *The Log of Bob Bartlett; The True Story of Forty Years of Seafaring and Exploration* (New York: G.P. Putnam's Sons, 1928) (hereafter cited as *The Log of Bob Bartlett*), pp. 96–97.
16 Lewis Amadeus Anspach, *A History of the Island of Newfoundland*, p. 443.
17 Chesley W. Sanger, "Newfoundland Seal Fishery," pp. 91–92; H.M. Mosdell, ed., *Chafe's Sealing Book*, p. 16.
18 Lewis Amadeus Anspach, *A History of the Island of Newfoundland*, p. 421.
19 H.M. Mosdell, ed., *Chafe's Sealing Book*, p. 39.
20 Joseph Beete Jukes, *Excursions in and about Newfoundland, during the Years 1839 and 1840* (London: John Murray, 1842), Vol. 1 (hereafter cited as *Excursions*), p. 251; Joseph Hatton and Moses Harvey, *Newfoundland*, p. 248.
21 H.M. Mosdell, ed., *Chafe's Sealing Book*, p. 17.
22 Lewis Amadeus Anspach, *A History of the Island of Newfoundland*, p. 415; Joseph Hatton and Moses Harvey, *Newfoundland*, p. 248.
23 Edmund W. Gosse, ed., *The Life of Philip Henry Gosse...., (London: Kegan Paul, Trench, Trübner, 1890) (hereafter cited as Philip Henry Gosse)*, p. 48.
24 Chesley W. Sanger, "Newfoundland Seal Fishery," p. 91.
25 Lewis Amadeus Anspach, *A History of the Island of Newfoundland*, p. 423.
26 Captain Robert Abram Bartlett, *The Log of Bob Bartlett*, p. 97.
27 Ibid.
28 Chesley W. Sanger, "Newfoundland Seal Fishery," p. 96; DNE, s.v. "rams."
29 Joseph Beete Jukes, *Excursions*, pp. 261–62.

30   Lewis Amadeus Anspach, *A History of the Island of Newfoundland*, pp. 421–22.

31   Chesley W. Sanger, "Newfoundland Seal Fishery," p. 104.

32   Joseph Beete Jukes, *Excursions*, p. 259.

33   George Allan England, *The Greatest Hunt in the World* (Montreal: Tundra Books, 1975) (hereafter cited as *The Greatest Hunt in the World*), p. 209. Originally published in 1924 by Doubleday, Page, Garden City, New York, as *Vikings of the Ice, Being the Log of a Tenderfoot on the Great Newfoundland Seal Hunt.*

34   Chesley W. Sanger, "Newfoundland Seal Fishery," p. 120.

35   DNE, s.v. "gaff."

36   Philip Tocque, *Kaleidoscope Echoes, being Historical, Philosophical, Scientific, and Theological Sketches....*, ed. Annie S. Tocque (Toronto: Hunter, Rose, 1893) (hereafter cited as *Kaleidoscope Echoes*), p. 197.

37   Chesley W. Sanger, "Newfoundland Seal Fishery," p. 124.

38   Joseph Beete Jukes, *Excursions*, pp. 290–91.

39   Ibid., pp. 274–75.

40   Major William Howe Greene, *Wooden Walls Among the Ice Floes*, p. 132.

41   Chesley W. Sanger, "Newfoundland Seal Fishery," p. 134.

42   Joseph Beete Jukes, *Excursions*, p. 260, p. 275; DNE, s.v. "sparables"; George Allan England, *The Greatest Hunt in the World*, p. 30.

43   Joseph Beete Jukes, *Excursions*, pp. 259–60.

44   Patrick William Browne, *Where the Fishers Go; The Story of Labrador* (New York: Cochrane, 1909) (hereafter cited as *Where the Fishers Go*), p. 179.

45   Joseph Beete Jukes, *Excursions*, pp. 284–85.

46   Chesley W. Sanger, "Newfoundland Seal Fishery," p. 140.

47   John Harvey, "The Newfoundland Seal Hunters," *The Canadian Magazine*, Vol. 16, No. 3 (Jan. 1901) (hereafter cited as "The Newfoundland Seal Hunters"), p. 197.

48   Joseph Hatton and Moses Harvey, *Newfoundland*, pp. 257–58.

49   Joseph Beete Jukes, *Excursions*, pp. 272–73.

50   Chesley W. Sanger, "Newfoundland Seal Fishery," p. 136.

51   Ibid., p. 171.

52   Ibid., p. 165.

53   Joseph Beete Jukes, *Excursions*, pp. 313–14.

54   Captain Abram Kean, *Old and Young Ahead, A Millionaire in Seals, being the Life History of Captain Abram Kean....* (London: Heath Cranton, 1935) (hereafter cited as *Old and Young Ahead*), p. 131.

55   Chesley W. Sanger, "Newfoundland Seal Fishery," p. 171.

56   Major William Howe Greene, *Wooden Walls Among the Ice Floes*, p. 59.

57   Philip Tocque, *Kaleidoscope Echoes*, p. 199.

58   H.M. Mosdell, ed., *Chafe's Sealing Book*, p. 41.

59   DNE, s.v. "growler."

60   Joseph Beete Jukes, *Excursions*, pp. 295–96.

61   Ibid., p. 251ff.

62   Ibid., p. 306.

63   Joseph Hatton and Moses Harvey, *Newfoundland*, pp. 255–56.

64   Joseph Beete Jukes, *Excursions*, p. 320.

65   Sir Richard Henry Bonnycastle, *Newfoundland in 1842*, p. 132; Lewis Amadeus Anspach, *A History of the Island of Newfoundland*, p. 428.

66 Joseph Hatton and Moses Harvey, *Newfoundland*, p. 259; Sir Richard Henry Bon-nycastle, *Newfoundland in 1842*, p. 132; Edmund W. Gosse, ed., *Philip Henry Gosse*, p. 49; Lewis Amadeus Anspach, *A History of the Island of Newfoundland*, pp. 424–26.

67 Sir Richard Henry Bonnycastle, *Newfoundland in 1842*, p. 162.

68 Edmund W. Gosse, ed., *Philip Henry Gosse*, p. 47.

69 Ibid., pp. 47–51.

70 H.M. Mosdell, ed., *Chafe's Sealing Book*, p. 33.

71 William Wilson, *Newfoundland and Its Missionaries....* (Cambridge, Mass.: Dakin and Metcalf, 1866), pp. 276–77.

72 Lewis Amadeus Anspach, *A History of the Island of Newfoundland*, p. 422.

73 Chesley W. Sanger, "Newfoundland Seal Fishery," p. 185.

74 Joseph Beete Jukes, *Excursions*, p. 259.

75 Cyril Byrne, "Some Comments on the Social Circumstances of Mummering in Conception Bay and St. John's in the Nineteenth Century," *Newfoundland Quarterly*, Vol. 77, No. 4 (Winter 1981–82), p. 6, n. 9.

76 Quoted in Chesley W. Sanger, "Newfoundland Seal Fishery," p. 185.

77 Lewis Amadeus Anspach, *A History of the Island of Newfoundland*, p. 423.

78 Chesley W. Sanger, "Newfoundland Seal Fishery," p. 103.

79 Eric W. Sager, "The Merchants of Water Street and Capital Investment in Newfoundland's Traditional Economy," in Lewis R. Fischer and Eric W. Sager, eds., *The Enterprising Canadians: Entrepreneurs and Economic Development in Eastern Canada, 1820–1914....* (St. John's: Memorial Univ. of Newfoundland, 1979) (hereafter cited as "The Merchants of Water Street"), p. 85.

The Seal Hunt, 1862–1939
An Overview

1 Chesley W. Sanger, "Influence of Scottish Whalemen," p. 237.

2 Gordon Jackson, *The British Whaling Trade* (London: Adam and Charles Black, 1978) (hereafter cited as *The British Whaling Trade*), p. 126.

3 Ibid., p. 129.

4 Ibid., p. 144.

5 Chesley W. Sanger, "Influence of Scottish Whalemen," p. 238ff.

6 These were the *Wolf* (Walter Grieve) and the *Bloodhound* (Baine, Johnston).

7 Joseph Hatton and Moses Harvey, *Newfoundland*, p. 249; Chesley W. Sanger, "Influence of Scottish Whalemen," p. 242; Eric W. Sager, "The Merchants of Water Street," p. 85.

8 Chesley W. Sanger, "Newfoundland Seal Fishery," p. 21.

9 Ibid., p. 112.

10 Ibid.

11 Newfoundland. Colonial Secretary's Office, *Census ... 1884* (St. John's: 1886).

12 Michael Staveley, "Population Dynamics in Newfoundland: The Regional Patterns," in John J. Mannion, ed., *The Peopling of Newfoundland: Essays in Historical Geography* (St. John's: Institute of Social and Economic Research, Memorial Univ. of Newfoundland, 1977), pp. 69–70.

13 Chesley W. Sanger, "Newfoundland Seal Fishery," p. 21; H.M. Mosdell, ed., *Chafe's Sealing Book*, p. 42.

14 Newfoundland. General Assembly. House of Assembly, *Journal of the House of Assembly ... 1891* through *Journal ... 1900*, Appendices.

15   David Alexander, "Newfoundland's Traditional Economy and Development to 1934," *Acadiensis*, Vol. 5, No. 2 (Spring 1976) (hereafter cited as "Traditional Economy," p. 63.

16   Chesley W. Sanger, "Newfoundland Seal Fishery," p. 112.

17   Daniel Woodley Prowse, *A History of Newfoundland*, p. 453.

18   Captain Robert Abram Bartlett, *The Log of Bob Bartlett*, p. 100.

19   Joseph Hatton and Moses Harvey, *Newfoundland*, p. 249; Chesley W. Sanger, "The Evolution of Sealing," p. 150.

20   Shannon Ryan, "The Newfoundland Cod Fishery," p. 11.

21   David Alexander, "Traditional Economy," p. 60.

22   William Barr, "The Role of Canadian and Newfoundland Ships in the Development of the Soviet Arctic," *Newfoundland Quarterly*, Vol. 73, No. 2 (Summer 1977), pp. 20–21; Newfoundland and Labrador. Provincial Archives (hereafter cited as PANL), GN1/10/0, 42/5/2, Morris to ———, 23 March 1917.

23   *Canadian Fisherman*, Vol. 3, No. 5 (May 1916), p. 177.

24   Ibid., No. 4 (Apr. 1916), p. 147.

25   E. Calvin Coish, *Season of the Seal: The International Storm Over Canada's Seal Hunt* (St. John's: Breakwater, 1979) (hereafter cited as *Season of the Seal*), p. 34; Farley Mowat, *Wake of the Great Sealers*, p. 138.

26   John Stacey Colman, "The Present State of the Newfoundland Seal Fishery," *Journal of Animal Ecology*, Vol. 6, No. 1 (May 1937) (hereafter cited as "The Present State of the Newfoundland Seal Fishery"), p. 153.

27   David Edwin Keir, *The Bowring Story* (London: Bodley Head, 1962) (hereafter cited as *The Bowring Story*), p. 158.

28   Ibid., p. 163.

29   Robert Brown Job, *John Job's Family: A Story of his Ancestors and Successors and their Business Connections with Newfoundland and Liverpool, 1730 to 1953* (St. John's: Telegram Printing, 1953) (hereafter cited as *John Job's Family*).

30   Thomas E. Appleton, *Usque Ad Mare; A History of the Canadian Coast Guard and Marine Services* (Ottawa: Dept. of Transport, 1968), p. 55.

31   Robert Brown Job, *John Job's Family*, p. 59.

32   David Alexander, "The Economic History of a Country and a Province," *Canadian Forum*, Vol. 53, No. 638 (Mar. 1974), p. 12.

33   David Alexander, "Development and Dependence in Newfoundland, 1880–1970," *Acadiensis*, Vol. 4, No. 1 (Autumn 1974), p. 20.

34   Ibid.

35   *Canadian Fisherman*, Vol. 2, No. 7 (July 1915), p. 216.

36   Ibid., Vol. 15, No. 4 (Apr. 1928), p. 116.

37   Captain Abram Kean, "Commentary on the Seal Hunt," in Joseph Roberts Smallwood, ed., *The Book of Newfoundland* (St. John's: Newfoundland Book Publishers, 1937–75), Vol. 1, p. 75.

38   Captain Abram Kean, "The Seal Fishery for 1934," *Newfoundland Quarterly 1901–1976: 75th Anniversary Edition* (St. John's: Creative Printers and Publishers, 1976, p. 142. Kean's article was originally published in July 1934.

39   *Canadian Fisherman*, Vol. 18, No. 2 (Feb. 1931), p. 43.

40   R.A. MacKay and S.A. Saunders, "Primary Industries," in Robert Alexander MacKay, ed., *Newfoundland; Economic, Diplomatic, and Strategic Studies* (Toronto: Oxford Univ. Press, 1946) (hereafter cited as "Primary Industries"), p. 88.

41 John Stacey Colman, "The Present State of the Newfoundland Seal Fishery,"
    pp. 145–59.

42 David Ernest Sergeant, "Harp Seals and the Sealing Industry," *Canadian Audubon*, Vol.
    25, No. 2 (Mar./Apr. 1963) (hereafter cited as "Harp Seals and the Sealing Industry"),
    p. 33.

43 Captain Robert Abram Bartlett, *The Log of Bob Bartlett*, p. 103.

44 *Canadian Fisherman*, Vol. 3, No. 5 (May 1916), p. 177.

45 Nicholas Smith, *Fifty-two Years at the Labrador Fishery* (London: Arthur H. Stockwell,
    1936) (hereafter cited as *Fifty-two Years at the Labrador Fishery*), p. 142; *Canadian
    Fisherman*, Vol. 1, No. 7 (July 1914), p. 195.

46 S.J.R. Noel, Politics in Newfoundland (Toronto: Univ. of Toronto Press, 1973),
    pp. 146–47.

47 *Canadian Fisherman*, Vol. 14, No. 2 (Feb. 1927), p. 43.

48 Ibid., Vol. 14, No. 6 (June 1927), p. 193.

49 Ibid., Vol. 15, No. 6 (June 1928), p. 193.

50 Ibid., Vol. 16, No. 3 (Mar. 1929), p. 39.

51 Ibid., Vol. 15, No. 6 (June 1928), p. 193.

52 From 1924 to 1941 inclusive, Levi Chafe published annual reports on the Newfoundland
    sealing fleet in the appendices to the House of Assembly journals. Although Chafe's
    reports were never published in a single volume, the Centre for Newfoundland Studies,
    Memorial University of Newfoundland, has a complete collection. The information on
    the *Sir William* is from the report for 1931. (Newfoundland. Memorial University, St.
    John's. Centre for Newfoundland Studies, "Reports of the Newfoundland Sealing Fleet,
    1924–1941," by Levi Chafe; further references to his reports will appear as CNS, Chafe,
    followed by the year of the report.)

53 Nova Scotia. General Assembly. House of Assembly, *Journal and Proceedings of the
    House of Assembly ... 1828* (Halifax: 1828), p. 285.

54 Thomas F. Knight, *Shore and Deep Sea Fisheries of Nova Scotia* (Halifax: Queen's
    Printer, 1867) (hereafter cited as *Shore and Deep Sea Fisheries of Nova Scotia*),
    pp. 34–35.

55 Michael Carroll, *The Seal and Herring Fisheries of Newfoundland. Together with a Con-
    densed History of the Island* (Montreal: John Lovell, 1873), p. 27.

56 Natashquan and Pointe-aux-Esquimaux boasted a combined sealing fleet of ap-
    proximately 40 schooners in the 1880s. See C.H. Farnham, "Labrador," *Harper's New
    Monthly Magazine*, Vol. 71, No. 425 (Oct. 1885) (hereafter cited as "Labrador"), p. 655.

57 Thomas F. Knight, *Shore and Deep Sea Fisheries of Nova Scotia*, p. 34.

58 James A. Farquhar, *Farquhar's Luck* (Halifax: Petheric Press, 1980), pp. 143–45.

59 H.M. Mosdell, ed., *Chafe's Sealing Book*, p. 77.

60 *Canadian Fisherman*, Vol. 3, No. 5 (May 1916), pp. 176–77; ibid., No. 6 (June 1916),
    p. 210.

61 *Evening Telegram* (St. John's), 9 Mar. 1929, p. 2.

62 Newfoundland. General Assembly. House of Assembly, *Journal of the House of Assemb-
    ly ... 1914*, Appendix, p. 514. I am grateful to Professor J.K. Hiller of Memorial Univer-
    sity of Newfoundland for this reference.

63 *Canadian Fisherman*, Vol. 3, No. 5 (May 1916), p. 177.

64 Ibid., No. 6 (June 1916), p. 210.

65 David Ernest Sergeant, "Exploitation and Conservation of Harp and Hood Seals," *Polar Record*, Vol. 12, No. 80 (May 1965) (hereafter cited as "Exploitation and Conservation of Harp and Hood Seals"), p. 542.

66 Newfoundland. Department of Natural Resources, *Report of the Newfoundland Fisheries Board and General Review of the Fisheries for the Years 1937 and 1938* (St. John's: 1940), p. 29.

67 Ibid., *Report of the Newfoundland Fisheries Board and General Review of the Fisheries for the Years 1939 and 1940* (St. John's: 1941), pp. 12–13.

68 CNS, Chafe, 1940.

The Vessel-Based Hunt
1862–1939

1 Lieutenant William Maxwell, "The Newfoundland Seal Fishery," *Nature*, Vol. 10 (Aug. 1874) (hereafter cited as "The Newfoundland Seal Fishery"), p. 264.

2 *Canadian Fisherman*, Vol. 16, No. 1 (Jan. 1929), p. 45.

3 R.A. MacKay and S.A. Saunders, "Primary Industries," p. 89; PANL, P5/11/6, Box 2, "Sealing Voyage Account, S.S. *Adventure*, 1908."

4 James E. Candow, "Preliminary Observations on the History of James Ryan Ltd.," manuscript on file (No. ARO-0014[i]), Atlantic Regional Office, Canadian Parks Service, Environment Canada, Halifax, 1981, p. 3.

5 Nicholas Smith, *Fifty-two Years at the Labrador Fishery*, p. 101.

6 Patrick William Browne, *Where the Fishers Go*, p. 179.

7 Kevin Major, "Terra Nova National Park 'Human History Study': A History of Southern Bonavista Bay from Alexander Bay to Goose Bay," Manuscript Report Series, No. 351, Environment Canada — Parks, Ottawa, 1979, p. 54; Chesley W. Sanger, "Newfoundland Seal Fishery," p. 212.

8 Chesley W. Sanger, "Newfoundland Seal Fishery," p. 206.

9 Major William Howe Greene, *Wooden Walls Among the Ice Floes*, pp. 268–69. Nominal horsepower is determined by a formula based on cylinder size and is not comparable to horsepower used for automobiles.

10 Sanger mistakenly gives 1896 as the purchase date of the *Merlin* and therefore underestimates the power of the early wooden walls. *See* Chesley W. Sanger, "Newfoundland Seal Fishery," p. 106.

11 Major William Howe Greene, *Wooden Walls Among the Ice Floes*, pp. 274–75.

12 Chesley W. Sanger, "Newfoundland Seal Fishery," p. 107; Major William Howe Greene, *Wooden Walls Among the Ice Floes*, p. 54.

13 Lieutenant William Maxwell, "The Newfoundland Seal Fishery," p. 264.

14 Newfoundland. Laws, Statutes, etc., *Acts of the General Assembly of Newfoundland* (St. John's: imprint varies) (hereafter cited as NLS), 36 Vict., cap. 9.

15 Ibid., 46 Vict., cap. 1.

16 Ibid., 50 Vict., cap. 23.

17 Ibid., 55 Vict., cap. 2.

18 Ibid., 4 Geo. V, cap. 19.

19 Cassie Brown, *Death on the Ice: The Great Newfoundland Sealing Disaster of 1914* (Toronto: Doubleday Canada, 1974) (hereafter cited as *Death on the Ice*), p. 14.

20 NLS, 6 Geo. V, cap. 24.

21 Ibid., 12 Geo. V, cap. 28.

22 Ibid., 15 Geo. V, cap. 2.

23 Ibid., 16 & 17 Geo. V, cap. 21.

24 Lieutenant William Maxwell, "The Newfoundland Seal Fishery," p. 265.

25 John Harvey, "The Newfoundland Seal Hunters," pp. 201–2.

26 Gordon Jackson, *The British Whaling Trade*, p. 149.

27 George Allan England, *The Greatest Hunt in the World*, pp. 299–301.

28 Major William Howe Greene, *Wooden Walls Among the Ice Floes*, p. 15.

29 Ibid., pp. 135–36.

30 Lieutenant William Maxwell, "The Newfoundland Seal Fishery," p. 264.

31 David Moore Lindsay, *A Voyage to the Arctic in the Whaler Aurora* (Boston: D. Estes, 1911) (hereafter cited as *A Voyage to the Arctic*), p. 46.

32 George Allan England, *The Greatest Hunt in the World*, pp. 242–43.

33 Chesley W. Sanger, "Newfoundland Seal Fishery," p. 109.

34 Major William Howe Greene, *Wooden Walls Among the Ice Floes*, pp. 196–97.

35 Chesley W. Sanger, "Newfoundland Seal Fishery," p. 99.

36 John Harvey, "The Newfoundland Seal Hunters," p. 203.

37 George Allan England, *The Greatest Hunt in the World*, p. 194.

38 David Moore Lindsay, *A Voyage to the Arctic*, p. 41.

39 Chesley W. Sanger, "Newfoundland Seal Fishery," p. 176.

40 NLS, 61 Vict., cap. 4.

41 Lieutenant William Maxwell, "The Newfoundland Seal Fishery," p. 265.

42 Chesley W. Sanger, "Newfoundland Seal Fishery," p. 120; John Harvey, "The Newfoundland Seal Hunters," p. 201; Major William Howe Greene, *Wooden Walls among the Ice Floes*, p. 113.

43 Cassie Brown, *Death on the Ice*, p. 34; Lieutenant William Maxwell, "The Newfoundland Seal Fishery," p. 265.

44 Chesley W. Sanger, "Newfoundland Seal Fishery," pp. 120–21.

45 Lieutenant William Maxwell, "The Newfoundland Seal Fishery," p. 265.

46 Chesley W. Sanger, "Newfoundland Seal Fishery," p.163.

47 Ibid., p. 171.

48 George Allan England, *The Greatest Hunt in the World*, p. 68; Major William Howe Greene, *Wooden Walls Among the Ice Floes*, p. 165; DNE, s.v. "ash-cat."

49 David Moore Lindsay, *A Voyage to the Arctic*, p. 51.

50 DNE, s.v. "pinnacle."

51 Lieutenant William Maxwell, "The Newfoundland Seal Fishery," p. 265.

52 Major William Howe Greene, *Wooden Walls Among the Ice Floes*, p. 165; DNE, s.v. "side stick."

53 Charles Ryle Fay, *Life and Labour in Newfoundland* (Toronto: Univ. of Toronto Press, 1956), p. 64.

54 George Allan England, *The Greatest Hunt in the World*, p. 76.

55 Chesley W. Sanger, "Newfoundland Seal Fishery," p. 209.

56 Ibid., p. 213.

57 Chesley W. Sanger, "Newfoundland Seal Fishery," p. 136.

58 Major William Howe Greene, *Wooden Walls Among the Ice Floes*, p. 132.

59 Chesley W. Sanger, "Newfoundland Seal Fishery," pp. 126–28.

60 Ibid.; DNE, s.v. "sculping knife."

61 George Allan England, *The Greatest Hunt in the World*, p. 61; Major William Howe Greene, *Wooden Walls Among the Ice Floes*, p. 166; Lieutenant William Maxwell, "The Newfoundland Seal Fishery," p. 265.

62 DNE, s.v. "dory."

63 John Gardner, *The Dory Book* (Camden, Maine: International Marine Publishing, 1978), p. 4.

64 Lieutenant William Maxwell, "The Newfoundland Seal Fishery," p. 265.

65 John Harvey, "The Newfoundland Seal Hunters," p. 203.

66 Ibid., p. 204; George Allan England, *The Greatest Hunt in the World*, pp. 295–97, 352.

67 DNE, s.v. "seal dog"; George Allan England, *The Greatest Hunt in the World*, p. 287.

68 Joseph Hatton and Moses Harvey, *Newfoundland,* p. 255.

69 Lieutenant William Maxwell, "The Newfoundland Seal Fishery," p. 265.

70 PANL, GN 2/5, 107-E, "Report of Commissioners. In the matter of the Enquiry respecting the disasters at the seal fishery in 1914," 24 Feb. 1915, pp. 15–16.

71 Lieutenant William Maxwell, "The Newfoundland Seal Fishery," p. 265.

72 Chesley W. Sanger, "Newfoundland Seal Fishery," p. 144.

73 Major William Howe Greene, *Wooden Walls Among the Ice Floes*, p. 216.

74 Cassie Brown, *Death on the Ice*, p. 65.

75 David Moore Lindsay, *A Voyage to the Arctic*, p. 50.

76 Major William Howe Greene, *Wooden Walls Among the Ice Floes*, p. 230.

77 Ibid., p. 227.

78 Chesley W. Sanger, "Newfoundland Seal Fishery," p. 155.

79 Lieutenant William Maxwell, "The Newfoundland Seal Fishery," p. 265.

80 Chesley W. Sanger, "Newfoundland Seal Fishery," p. 134.

81 Major William Howe Greene, *Wooden Walls Among the Ice Floes*, p. 219.

82 Ibid., p. 216.

83 Ibid., p. 225; Cassie Brown, *Death on the Ice*, p. 35; George Allan England, *The Greatest Hunt in the World*, pp. 111–12.

84 Major William Howe Greene, *Wooden Walls Among the Ice Floes*, p. 226; George Allan England, *The Greatest Hunt in the World*, p. 198; Lieutenant William Maxwell, "The Newfoundland Seal Fishery," p. 265.

85 Major William Howe Greene, *Wooden Walls Among the Ice Floes*, p. 191.

86 Lieutenant William Maxwell, "The Newfoundland Seal Fishery," p. 265.

87 George Allan England, *The Greatest Hunt in the World*, p. 197.

88 Cassie Brown, *Death on the Ice*, p. 35.

89 Joseph Hatton and Moses Harvey, *Newfoundland,* p. 256.

90 George Allan England, *The Greatest Hunt in the World*, pp. 299–301.

91 Ibid., p. 297; DNE, s.v. "nunch-bag."

92 John Harvey, "The Newfoundland Seal Hunters," p. 200.

93 Ibid., p. 204.

94 George Allan England, *The Greatest Hunt in the World*, p. 287.

95 Lieutenant William Maxwell, "The Newfoundland Seal Fishery," pp. 265–66.

96 George Allan England, *The Greatest Hunt in the World*, pp. 345–48.

97 PANL, GN 2/5, 376-B, Bowring to Squires, 12 Jan. 1921; ibid., Campbell to Squires, 20 Apr. 1921.

98 Frank H. Ellis, *Canada's Flying Heritage* (Toronto: Univ. of Toronto Press, 1980) (hereafter cited as *Canada's Flying Heritage*), p. 215.

99   Sidney Cotton, *Aviator Extraordinary: The Sidney Cotton Story; As Told to Ralph Barker* (London: Chatto and Windus, 1969) (hereafter cited as *Aviator Extraordinary*), pp. 58–60.

100   PANL, GN 2/5, 376-B, Job Brothers to Squires, 29 Apr. 1922.

101   E.L. Chicanot, "New 'Eyes' for the Sealing Fleet," *Scientific American*, Vol. 138, No. 5 (May 1928) (hereafter cited as "New 'Eyes' for the Sealing Fleet"), p. 410; Frank H. Ellis, *Canada's Flying Heritage*, p. 216.

102   Sidney Cotton, *Aviator Extraordinary*, p. 65.

103   Frank H. Ellis, *Canada's Flying Heritage*, p. 216.

104   *Evening Telegram* (St. John's), 19 Apr. 1923, p. 8.

105   Frank H. Ellis, *Canada's Flying Heritage*, p. 216.

106   *Canadian Fisherman*, Vol. 15, No. 3 (Mar. 1928), p. 99.

107   E.L. Chicanot, "New 'Eyes' for the Sealing Fleet," p. 411.

108   *Evening Telegram* (St. John's), 18 Mar. 1930, p. 6.

109   Captain Abram Kean, *Old and Young Ahead*, p. 142.

110   George Allan England, *The Greatest Hunt in the World*, p. 225.

111   H.M. Mosdell, ed., *Chafe's Sealing Book*, p. 28.

112   Ibid.

113   PANL, GN 2/5, 376-B, Job Brothers to colonial secretary, 25 May 1922.

114   Chesley W. Sanger, "Newfoundland Seal Fishery," p. 152.

115   Sunset occurred at 6:09 p.m. on 26 March 1888. By 5 April the sun was setting at 6:24 p.m. *A Year Book and Almanac of Newfoundland, for 1888....* (St. John's: Queen's Printer, 1888), pp. 6–7.

116   Major William Howe Greene, *Wooden Walls Among the Ice Floes*, pp. 63–64.

117   Cassie Brown, *Death on the Ice*, pp. 37–38.

118   H.M. Mosdell, ed., *Chafe's Sealing Book*, p. 43.

119   George Allan England, *The Greatest Hunt in the World*, p. 63.

120   Cassie Brown, *Death on the Ice*, p. 200; Major William Howe Greene, *Wooden Walls Among the Ice Floes*, p. 64.

121   Farley Mowat, *Wake of the Great Sealers*, p. 45.

122   Chesley W. Sanger, "Newfoundland Seal Fishery," p. 181; Joseph Hatton and Moses Harvey, *Newfoundland*, p. 255; Major William Howe Greene, *Wooden Walls Among the Ice Floes*, p. 100.

123   George Allan England, *The Greatest Hunt in the World*, p. 74.

124   Joseph Hatton and Moses Harvey, *Newfoundland*, p. 255; DNE, s.v. "duff."

125   DNE, s.v. "bogie."

126   George Allan England, *The Greatest Hunt in the World*, p. 101; David Moore Lindsay, *A Voyage to the Arctic*, p. 56.

127   Cassie Brown, *Death on the Ice*, p. 36.

128   H.M. Mosdell, ed., *Chafe's Sealing Book*, p. 43.

129   PANL, GN2/5, 107-D, Campbell to Bennett, 20 Mar. 1915.

130   Ibid., 107-B, Brehm to Watson, 24 Feb. 1911.

131   H.M. Mosdell, ed., *Chafe's Sealing Book*, p. 44.

132   Captain Robert Abram Bartlett, *The Log of Bob Bartlett*, pp. 40–41.

133   Cassie Brown, *Death on the Ice*, p. 23.

134   Lieutenant William Maxwell, "The Newfoundland Seal Fishery," p. 265.

135   PANL, GN2/5, 107-E, Johnson to Davidson, 27 Feb. 1915.

136 Cassie Brown, *Death on the Ice*, p. 261.
137 PANL, GN2/5, 107-E, Johnson to Davidson, 27 Feb. 1915.
138 NLS, 56 Vict., cap. 22.
139 *Evening Telegram* (St. John's), 26 May 1893, p. 3.
140 NLS, 4 Geo. V, cap. 19.
141 Cassie Brown, *Death on the Ice*, p. 31.
142 NLS, 5 Geo. V, cap. 6, War Session.
143 Ibid., 6 Geo. V, cap. 24.
144 Ibid., 8 Edward VII, cap. 5.
145 Ibid., 11 Geo. V, cap. 28.
146 George Allan England, *The Greatest Hunt in the World*, p. 76.
147 Ibid., pp. 204–10.
148 "Shipwrecked at the Front," *Decks Awash*, Vol. 13, No. 3 (May/June 1984), pp. 50–51.
149 Clayton L. King, "The *Viking*'s Last Cruise," in Joseph Roberts Smallwood, ed., *The Book of Newfoundland* (St. John's: Newfoundland Book Publishers, 1937–75), Vol. 1, pp. 76–85.
150 Captain Robert Abram Bartlett, "The Sealing Saga of Newfoundland," *National Geographic Magazine*, Vol. 56, No. 1 (July 1929) (hereafter cited as "The Sealing Saga of Newfoundland"), pp. 122–23.
151 Ibid., pp. 124–25.
152 PANL, GN 2/5, 107-E, "Report of Commissioners. In the matter of the Enquiry respecting the disasters at the seal fishery of 1914," 24 Feb. 1915, p. 15.
153 Samuel George Archibald, Some Account of the Seal Fishery of Newfoundland and the Mode of Preparing Seal Oil.... (Edinburgh: Murray and Gibb, 1852), p. 6.
154 James J. Fogarty, "The Seal-Skinners' Union," in Joseph Roberts Smallwood, ed., *The Book of Newfoundland* (St. John's: Newfoundland Book Publishers, 1937–75) (hereafter cited as "The Seal-Skinners' Union"), Vol. 2, p. 100.
155 Briton Cooper Busch, *The War against the Seals: A History of the North American Seal Fishery* (Kingston: McGill-Queen's University Press, 1985), p. 63.
156 Joseph Hatton and Moses Harvey, *Newfoundland*, pp. 258–59; *see also* J.K. Hiller, "The Newfoundland Seal Fishery: An Historical Introduction," *Bulletin of Canadian Studies*, Vol. 7, No. 2 (Winter 1983/84), p. 50.
157 Lieutenant William Maxwell, "The Newfoundland Seal Fishery," p. 266.
158 CNS, Chafe, 1930.
159 *Canadian Fisherman*, Vol. 13, No. 3 (Mar. 1926), p. 106.
160 James J. Fogarty, "The Seal-Skinners' Union," p. 100.
161 Captain Robert Abram Bartlett, "The Sealing Saga of Newfoundland," p. 130.
162 *Canadian Fisherman*, Vol. 15, No. 1 (Jan. 1928), p. 20.
163 *Evening Telegram* (St. John's), 22 Feb. 1929, p. 6.
164 *Canadian Fisherman*, Vol. 15, No. 6 (June 1928), p. 193.
165 Chesley W. Sanger, "Newfoundland Seal Fishery," pp. 187–90.
166 W.A. Black, "The Labrador Floater Codfishery," *Annals of the Association of American Geographers*, Vol. 50 (Sept. 1960), pp. 269–70.
167 Patrick William Browne, *Where the Fishers Go*, p. 179; George Allan England, *The Greatest Hunt in the World*, p. 34.
168 H.M. Mosdell, ed., *Chafe's Sealing Book*, p. 38; George Allan England, *The Greatest Hunt in the World*, p. 34.

169 H.M. Mosdell, ed., *Chafe's Sealing Book*, p. 38; Paul O'Neill, *A Seaport Legacy: The Story of St. John's, Newfoundland* (Erin, Ont.: Press Porcepic, 1976), Vol. 2, pp. 973–74; Briton Cooper Busch, "The Newfoundland Sealers' Strike of 1902," *Labour/Le Travail*, No. 14 (Fall 1984), pp. 73–101.

170 H.M. Mosdell, ed., *Chafe's Sealing Book*, p. 36.

171 George Allan England, *The Greatest Hunt in the World*, p. 77.

172 Lieutenant William Maxwell, "The Newfoundland Seal Fishery," p. 265; Captain Robert Abram Bartlett, "The Sealing Saga of Newfoundland," p. 130.

173 H.M. Mosdell, ed., *Chafe's Sealing Book*, p. 44.

174 PANL, P5/11/6, Box 2, "Sealing Voyage Account, S.S. *Adventure*, 1908."

175 Cater Winsor, "Reminiscences of Rev. Cater Winsor," in Naboth Winsor, ed., *"By Their Works": A History of the Wesleyville Congregation: Methodist Church 1874–1925, United Church 1925–1974* (n.p.: privately published, 1976), p. 47.

176 Captain Abram Kean, *Old and Young Ahead*, p. 46; Robert Brown Job, *John Job's Family*, pp. 87–88.

177 Figures are based on Table 1 in John Stacey Colman, "The Newfoundland Seal Fishery and the Second World War," *Journal of Animal Ecology*, Vol. 18, No. 1 (May 1949) (hereafter cited as "The Newfoundland Seal Fishery and the Second World War"), p. 43.

178 Captain Robert Abram Bartlett, "The Sealing Saga of Newfoundland," p. 123.

179 *Canadian Fisherman*, Vol. 16, No. 1 (Jan. 1928), p. 20.

180 George Allan England, *The Greatest Hunt in the World*, p. 226.

181 *Canadian Fisherman*, Vol. 15, No. 4 (Apr. 1928), p. 116.

182 George Allan England, *The Greatest Hunt in the World*, pp. 77–78.

183 Great Britain. Newfoundland Royal Commission, *Newfoundland Royal Commission, 1933: Report* (London: HMSO, 1934), p. 79.

184 Chesley W. Sanger, "Newfoundland Seal Fishery," p. 195.

185 John Roper Scott, "The Function of Folklore in the Interrelationship of the Newfoundland Seal Fishery and the Home Communities of the Sealers," MA thesis, Memorial Univ. of Newfoundland, St. John's, 1974 (hereafter cited as "The Function of Folklore"), pp. 233–39.

186 Captain Robert Abram Bartlett, *The Log of Bob Bartlett*, p. 257.

187 John Roper Scott, "The Function of Folklore," p. 145.

188 Article, no title, *Public Ledger and Newfoundland General Advertiser* (St. John's), 14 Feb. 1832, p. 3.

189 George Allan England, *The Greatest Hunt in the World*, p. 77.

The Seal Hunt Since 1939
An Overview

1 Newfoundland. Department of Natural Resources, *Report of the Newfoundland Fisheries Board and General Review of the Fisheries for the Year 1945 with Statistical Survey* (St. John's: 1947), p. 16.

2 G.S. Watts, "The Impact of the War," in Robert Alexander MacKay, ed., *Newfoundland; Economic, Diplomatic, and Strategic Studies* (Toronto: Oxford Univ. Press, 1946), p. 221.

3 J.S. Colman, "The Newfoundland Seal Fishery and the Second World War," p. 45.

4 *Trade News*, Vol. 4, No. 4 (Oct. 1951), p. 3; John Parker, *Newfoundland: 10th Province of Canada* (London: Lincolns-Prager, 1950), p. 57.

5 John Stacey Colman, "The Newfoundland Seal Fishery and the Second World War," p. 45.

6 *Trade News*, Vol. 10, No. 8 (Feb. 1958), p. 3.

7 *Decks Awash*, Vol. 7, No. 1 (Feb. 1978), p. 48; Captain Morrissey Johnson, "Thirty Years at the Seal Hunt," pp. 41–42.

8 Richard Gwyn, *Smallwood, The Unlikely Revolutionary* (Toronto: McClelland and Stewart, 1968) (hereafter cited as *Smallwood*), p. 124.

9 E. Pazdzior, "The Fishing Industry," in Ian McAllister, ed., *Newfoundland and Labrador: The First Fifteen Years of Confederation* (St. John's: Dicks, 1966) (hereafter cited as "The Fishing Industry"), p. 131.

10 Richard Gwyn, *Smallwood*, p. 168.

11 Ibid., p. 172.

12 E. Pazdzior, "The Fishing Industry," p. 125.

13 David Alexander, *The Decay of Trade: An Economic History of the Newfoundland Saltfish Trade, 1935–1965* (St. John's: Institute of Social and Economic Research, Memorial Univ. of Newfoundland, 1977), pp. 4–16.

14 Ibid., p. 141.

15 *Trade News*, Vol. 10, No. 8 (Feb. 1958), p. 3.

16 Ibid., Vol. 4, No. 4 (Oct. 1951), p. 3.

17 Ibid., Vol. 6, No. 9 (Mar. 1954), p. 3.

18 Ibid., Vol. 18, No. 1 (July 1965), pp. 7–8.

19 David Ernest Sergeant, "Harp Seals and the Sealing Industry," p. 33.

20 *Trade News*, Vol. 4, No. 4 (Oct. 1951), p. 3; ibid., Vol. 11, No. 10 (Apr. 1959), p. 3.

21 *Evening Telegram* (St. John's), 3 Mar. 1961, p. 4.

22 Ibid., 3 Mar. 1967, p. 6.

23 Ibid., 1 Mar. 1968, p. 26.

24 *Trade News*, Vol. 18, No. 1 (July 1965), p. 7.

25 *Evening Telegram* (St. John's), 3 Mar. 1967, p. 6.

26 David Ernest Sergeant, "Harp Seals and the Sealing Industry," p. 33.

27 Ibid.

28 H.D. Fisher, "Utilization of Atlantic Harp Seal Populations," in North American Wildlife Conference, *Transactions of the Twentieth North American Wildlife Conference* (Washington, D.C.: Wildlife Management Institute, 1955) (hereafter cited as "Utilization of Atlantic Harp Seal Populations"), p. 510.

29 *Trade News*, Vol. 2, No. 6 (Dec. 1949), p. 4.

30 H.D. Fisher, "Utilization of Atlantic Harp Seal Populations," pp. 512–13.

31 Douglas H. Pimlott, "Seals and Sealing in the North Atlantic," *Canadian Audubon*, Vol. 28, No. 2 (Mar./Apr. 1966) (hereafter cited as "Seals and Sealing in the North Atlantic"), p. 38.

32 *Trade News*, Vol. 7, No. 7 (Jan. 1955), p. 17.

33 David Ernest Sergeant, "Exploitation and Conservation of Harp and Hood Seals," p. 547; David Ernest Sergeant, *Studies in the Sustainable Catch of Harp Seals in the Western North Atlantic* (Ste-Anne-de-Bellevue, Que.: Fisheries Research Board of Canada, Arctic Biological Station, 1959) (hereafter cited as *Sustainable Catch of Harp Seals*).

34 David Ernest Sergeant, "Harp Seals and the Sealing Industry," p. 34.

35 Harold Horwood, "Tragedy on the Whelping Ice," *Canadian Audubon*, Vol. 22, No. 2 (Mar./Apr. 1960), pp. 37–41.

36 Harry R. Lillie, *The Path Through Penguin City* (London: Ernest Benn, 1955), p. 244.
37 Ibid., pp. 244–45.
38 Douglas H. Pimlott, "Seals and Sealing in the North Atlantic," p. 34.
39 Canada. Laws, Statutes, etc., *Seal Protection Regulations*, Statutory Order Regulations 64/443 (Order in Council P.C. 1964/1663).
40 "Review," *Canadian Audubon*, Vol. 28, No. 2 (Mar./Apr. 1966), p. 56.
41 "New Regulations to Protect Seals," *Canadian Audubon*, Vol. 26, No. 5 (Nov./Dec. 1964), p. 166.
42 Douglas H. Pimlott, "Seals and Sealing in the North Atlantic," pp. 35–36.
43 Brian Davies, *Savage Luxury: The Slaughter of the Baby Seals* (Toronto: Ryerson Press, 1970) (hereafter cited as *Savage Luxury*), p. 19.
44 Ibid., p. 30
45 Ibid., p. 87.
46 Ibid., p. 145.
47 Ibid., pp. 209–14.
48 *Decks Awash*, Vol. 7, No. 7 (Feb. 1978), pp. 55–56.
49 *Evening Telegram* (St. John's), 7 Mar. 1967, p. 3.
50 Douglas H. Pimlott, "The 1967 Seal Hunt," *Canadian Audubon*, Vol. 29, No. 2 (Mar./Apr. 1967) (hereafter cited as "The 1967 Seal Hunt"), p. 42.
51 David Ernest Sergeant, "Exploitation and Conservation of Harp and Hood Seals," p. 547.
52 Douglas H. Pimlott, "The 1967 Seal Hunt," p. 70.
53 "Norway Joins Canada in Ban on Hunting 'Whitecoat' Seals," *Fisheries of Canada*, Vol. 22, No. 8 (Feb./Mar. 1970), p. 3.
54 Brian Davies, *Savage Luxury*, pp. 200–201.
55 *Globe and Mail* (Toronto), 10 Feb. 1979, p. 2.
56 *Evening Telegram* (St. John's), 8 Mar. 1971, p. 2.
57 E. Calvin Coish, *Season of the Seal*, pp. 118–19.
58 *Globe and Mail* (Toronto), 10 Feb. 1979, p. 2.
59 *Evening Telegram* (St. John's), 12 Mar. 1973, p. 1.
60 David M. Lavigne, "Counting Harp Seals with Ultra-violet Photography," *Polar Record*, Vol. 18, No. 114 (Sept. 1976) (hereafter cited as "Counting Harp Seals"), p. 270.
61 David M. Lavigne, "Life or Death for the Harp Seal," *National Geographic*, Vol. 149, No. 1 (Jan. 1976), pp. 129–42.
62 David M. Lavigne, "Counting Harp Seals," p. 276.
63 E. Calvin Coish, *Season of the Seal*, p. 135.
64 *Evening Telegram* (St. John's), 3 Mar. 1976, p. 2.
65 Ibid., 10 Mar. 1976, p. 1.
66 International Commission for the Northwest Atlantic Fisheries, "Report of Scientific Advisors to Panel A (Seals)" in *ICNAF Redbook* (Dartmouth, N.S.: ICNAF, 1977), p. 96.
67 Staffan Soederberg and Lennart Almkvist, "In Prospect of the Seal Hunt in Canada 1977: A Comment" (Stockholm: Swedish Museum of Natural History, 1977), pp. 3–4.
68 *Evening Telegram* (St. John's), 16 Mar. 1977, p. 1.
69 E. Calvin Coish, *Season of the Seal*, pp. 140–45.
70 D.L. Dunn, *Canada's East Coast Sealing Industry 1976: A Socio-economic Review*, (Ottawa: Dept. of Fisheries and the Environment, 1977) (hereafter cited as *Canada's East Coast Sealing Industry 1976*), p. 6.
71 *Decks Awash*, Vol. 7, No. 1 (Feb. 1978), pp. 24–25.

Notes to pages 125–135

72 Cynthia Lamson, *"Bloody Decks and a Bumper Crop": The Rhetoric of Sealing Counter-protest* (St. John's: Institute of Social and Economic Research, Memorial Univ. of Newfoundland, 1979) (hereafter cited as *"Bloody Decks and a Bumper Crop"*), pp. 3–4.
73 D.L. Dunn, *Canada's East Coast Sealing Industry 1976*, p. 4.
74 *Decks Awash*, Vol. 7, No. 1 (Feb. 1978), pp. 28.
75 Captain Morrissey Johnson, "Thirty Years at the Seal Hunt," p. 44.
76 Cynthia Lamson, *"Bloody Decks and a Bumper Crop,"* pp. 6–7.
77 D.L. Dunn, *Canada's East Coast Sealing Industry 1976*, p. 24.
78 *Decks Awash*, Vol. 7, No. 1 (Feb. 1978), p. 10.
79 Quoted in E. Calvin Coish, *Season of the Seal*, p. 130.
80 Malcolm C. Mercer, *The Seal Hunt* (Ottawa: Dept. of Fisheries and Environment, 1976), p. 10.
81 G.H. Winters, "Production, Mortality, and Sustainable Yield of Northwest Atlantic Harp Seals (*Pagophilus groenlandicus*)," *Journal of the Fisheries Research Board of Canada*, Vol. 35, No. 9 (Sept. 1978), p. 1260.
82 Figure quoted in David M. Lavigne, "The Harp Seal Controversy Reconsidered," *Queen's Quarterly*, Vol. 85, No. 3 (Autumn 1978), p. 381.
83 Ibid., p. 382.
84 Quoted in *Decks Awash*, Vol. 7, No. 1 (Feb. 1978), p. 54.
85 *Evening Telegram* (St. John's), 1 Mar. 1977, p. 6.
86 *Globe and Mail* (Toronto), 10 Feb. 1979, p. 1.
87 Ibid., p. 2.
88 *Evening Telegram* (St. John's), 18 Feb. 1978, p. 1.
89 Ibid., 21 Feb. 1978, p. 1.
90 *Globe and Mail* (Toronto), 10 Feb. 1979, p. 2.
91 *Evening Telegram* (St. John's), 18 Feb. 1978, p. 1.
92 Ibid., 1 Mar. 1978, p. 2.
93 Ibid., 4 Mar. 1978, p. 6.
94 Ibid.
95 Ibid., 1 Mar. 1978, p. 2.
96 Ibid., 9 Mar. 1978, p. 18.
97 D.L. Dunn, *Canada's East Coast Sealing Industry 1976*.
98 D.M. Lavigne et al., *The 1977 Census of Western Atlantic Harp Seals Pagophilus groenlandicus* (Dartmouth, N.S.: Dept. of Fisheries and the Environment/Canadian Atlantic Fisheries Scientific Adivsory Committee, 1977).
99 *Evening Telegram* (St. John's), 29 Mar. 1978, p. 8. For the report in question, *see* Harry C. Rowsell, "1977 Sealing Activities of Newfoundland Landsmen and Ships on the Front: A Report to the Committee on Seals and Sealing and the Canadian Federation of Humane Societies" (1977), Ottawa.
100 "Enormous Boycott Predicted," *Mail-Star* (Halifax), 10 Jan. 1979, p. 10.
101 *Evening Telegram* (St. John's), 2 Mar. 1979, p. 6.
102 Ibid., 3 Mar. 1979, p. 3.
103 Ibid., 15 Mar. 1980, p. 2.
104 *Decks Awash*, Vol. 7, No. 1 (Feb. 1978), p. 54.
105 *Evening Telegram* (St. John's), 3 Mar. 1983, p. 2.
106 Ibid., 17 Mar. 1983, p. 4.
107 Ibid., p. 1.

206

108 *Globe and Mail* (Toronto), 12 Mar. 1982, p. 1.

109 Ibid., 8 Mar. 1984, p. 3; 21 Mar. 1983, p. 8.

110 Ibid., 26 Nov. 1983, p. 6.

111 "Anti-Fish Campaign Has 'Bogus Premise,'" *Chronicle-Herald* (Halifax), 29 Feb. 1984, p. 39.

112 C.H. Farnham, "Labrador," p. 654.

113 Chesley W. Sanger, "The Evolution of Sealing," pp. 142–45.

114 Wilfred Thomason Grenfell, *A Laborador Doctor: The Autobiography of Wilfred Thomason Grenfell* (London: Hodder and Stoughton, [1920]), p. 121.

115 CNS, Chafe, 1924.

116 Ibid., 1931.

117 *Evening Telegram* (St. John's), 29 Apr. 1899, p. 4.

118 *Decks Awash*, Vol. 7, No. 1 (Feb. 1978), p. 48.

119 *Canadian Fisherman*, Vol. 14, No. 6 (June 1927), p. 193.

120 *Trade News*, Vol. 4 (Oct. 1951), p. 3.

121 David Ernest Sergeant, *Sustainable Catch of Harp Seals*, p. 3.

122 *Canadian Fisheries Annual*, 1957, p. 69.

123 *Trade News*, Vol. 8, No. 9 (Mar. 1956), p. 10; *Evening Telegram* (St. John's), 1 Mar. 1951, p. 3.

124 Newfoundland. Memorial University, St. John's. Centre for Newfoundland Studies, "Review of the Newfoundland Fisheries for 1959," by the Honourable J.T. Cheeseman, Minister of Fisheries, p. 1.

125 Newfoundland. Department of Natural Resources. Fishery Research Commission, *Report No. 2* (St. John's: 1963), n.p.

126 Ibid.

127 Newfoundland. Statistics Agency, *Historical Statistics of Newfoundland and Labrador* (St. John's: Division of Printing Services, Dept. of Public Works and Services, 1977–), Vol. 2, No. 1, Table K-8.

128 *Decks Awash*, Vol. 7, No. 1 (Feb. 1978), p. 13.

129 D.L. Dunn, *Canada's East Coast Sealing Industry 1976*, p. 25.

130 *Decks Awash*, Vol. 7, No. 1 (Feb. 1978), pp. 10–24.

131 Newfoundland. Royal Commission on Labrador, *Report of the Royal Commission on Labrador* (St. John's: 1974), Vol. 3, p. 568.

132 Canadian Sealers' Association, *Annual Report ... 1983/84* (St. John's: 1984), p. 11.

The Vessel-Based Hunt
Since 1939

1 Interview, Captain Morrissey Johnson, 6 June 1984.

2 *Evening Telegram* (St. John's), 6 Mar. 1982, p. 3.

3 Captain Morrissey Johnson, "Thirty Years at the Seal Hunt," pp. 41–42.

4 John Stacey Colman, "The Newfoundland Seal Fishery and the Second World War," p. 40.

5 *Trade News*, Vcl. 4, No. 4 (Oct. 1951), p. 3.

6 Ibid., Vol. 7, No. 8 (Feb. 1955), p. 12.

7 Ibid., Vol. 6, No. 8 (Feb. 1954), p. 5.

8 David Edwin Keir, *The Bowring Story*, p. 404.

9 *Trade News*, Vol. 18, No. 1 (July 1965), p. 7; *Evening Telegram* (St. John's), 1 Mar. 1968, p. 26; pers. com., Captain Alfred M. Shaw, 4 Aug. 1984.

10 *Trade News*, Vol. 4, No. 4 (Oct. 1951), p. 3.

11 Interview, Captain Alfred M. Shaw, 25 July 1984.

12 David Edwin Keir, *The Bowring Story*, p. 404.

13 *Evening Telegram* (St. John's), 1 Mar. 1977, p. 1.

14 *Globe and Mail* (Toronto), 29 Mar. 1983, p. 10.

15 *Canadian Fisherman*, Vol. 12, No. 5 (May 1925), p. 16.

16 *Trade News*, Vol. 1, No. 5 (Nov. 1948), p. 16.

17 *Evening Telegram* (St. John's), 1 Mar. 1952, p. 4.

18 Guy David Wright, *Sons and Seals: A Voyage to the Ice* (St. John's: Institute of Social and Economic Research, Memorial Univ. of Newfoundland, 1984) (hereafter cited as *Sons and Seals*), p. 50.

19 H.J. Percey, *Merchant Ship Construction* (Glasgow: Brownside and Ferguson, 1983), pp. 143–45.

20 *Evening Telegram* (St. John's), 7 Mar. 1955, p. 3; *Trade News*, Vol. 10, No. 2 (Aug. 1957), p. 13.

21 *Evening Telegram* (St. John's), 3 Mar. 1964, p. 3.

22 Ibid., 5 Mar. 1965, p. 4.

23 Ibid., 12 Mar. 1970, p. 3.

24 Ibid., 6 Mar. 1982, p. 3.

25 Ibid., 6 Mar. 1974, p. 3. The St. John's press regularly identified Shaw Steamship Co. Ltd. of Halifax as owner of the *Arctic Endeavour*. Mayhaven Shipping was in fact a subsidiary formed by Captain William Alfred Shaw in 1947. Pers. com., Captain Alfred M. Shaw, 4 Aug. 1984.

26 *Evening Telegram* (St. John's), 7 Mar. 1977, p. 3.

27 Captain Morrissey Johnson interview, 6 June 1984; Captain Morrissey Johnson, "Thirty Years at the Seal Hunt," p. 43.

28 Interview, Harold Henriksen, 2 May 1984.

29 D.L. Dunn, *Canada's East Coast Sealing Industry 1976*, p. 44; Captain Morrissey Johnson interview, 6 June 1984; Harold Henriksen interview, 2 May 1984.

30 Captain Morrissey Johnson, "Thirty Years at the Seal Hunt," p. 43.

31 Ibid., p. 41.

32 *Decks Awash*, Vol. 7, No. 1 (Feb. 1978), p. 11.

33 Captain Morrissey Johnson interview, 6 June 1984.

34 *Evening Telegram* (St. John's), 4 Mar. 1969, p. 3.

35 Ibid., 4 Mar. 1972, p. 4.

36 Interview, Captain Alfred M. Shaw, 25 July 1984.

37 Guy David Wright, *Sons and Seals*, p. 56.

38 DNE, s.v. "creeper"; Captain Morrissey Johnson interview, 6 June 1984.

39 Brian Davies, *Savage Luxury*, p. 84.

40 H.D. Fisher, "Utilization of Atlantic Harp Seal Populations," p. 509.

41 Guy David Wright, *Sons and Seals*, p. 55ff.

42 Captain Morrissey Johnson interview, 6 June 1984.

43 Guy David Wright, *Sons and Seals*, p. 59.

44 Captain Morrissey Johnson interview, 6 June 1984.

45 Harold Henriksen interview, 2 May 1984.

46 D.L. Dunn, *Canada's East Coast Sealing Industry 1976*, p. 8.
47 Pers. com., Harold Henriksen, 27 Sept. 1984.
48 Guy David Wright, *Sons and Seals*, pp. 74–75. Wright gives the erroneous impression that this was the norm instead of the exception.
49 Brian Davies, *Savage Luxury*, p. 87.
50 *Evening Telegram* (St. John's), 6 Mar. 1982, p. 3.
51 Pers. com., Harold Henriksen, 11 May 1984; interview, Royal Cooper, 27 Aug. 1984.
52 Royal Cooper interview, 27 Aug. 1984.53 Ibid.
54 Harry R. Lillie, *The Path through Penguin City* (London: Ernest Benn, 1955), p. 243.
55 *Trade News*, Vol. 6, No. 9 (Mar. 1954), p. 5.
56 Guy David Wright, *Sons and Seals*, p. 56.
57 Newfoundland. Laws, Statutes, etc., *Worker's Compensation Legislation*, Statutes of Newfoundland, 1948, No. 30.
58 Pers. com., H.M. Andrews, 1 Nov. 1984.
59 D.L. Dunn, *Canada's East Coast Sealing Industry 1976*, p. 15.
60 Newfoundland. Laws, Statutes etc., *Worker's Compensation Fishing Regulations, 1980*, Newfoundland Regulation 277/80.
61 *Evening Telegram* (St. John's), 6 Mar. 1954, p. 3.
62 *Trade News*, Vol. 6, No. 8 (Feb. 1954), p. 8; Captain Morrissey Johnson, "Thirty Years at the Seal Hunt," pp. 43–44.
63 *Evening Telegram* (St. John's), 1 Mar. 1967, p. 3.
64 Guy David Wright, *Sons and Seals*, pp. 32–39.
65 Ibid., p. 39.
66 Ibid., p. 56.
67 *Evening Telegram* (St. John's), 19 Mar. 1980, p. 4.
68 Newfoundland. Department of Natural Resources, *Report of the Newfoundland Fisheries Board and General Review of the Fisheries for the Year 1945, with Statistical Survey* (St. John's: 1947), p. 16. All other figures in this paragraph are from Newfoundland Fisheries Board reviews.
69 *Trade News*, Vol. 1, No. 5 (Nov. 1948), p. 17.
70 H.D. Fisher, "Utilization of Atlantic Harp Seal Populations," p. 511.
71 *Trade News*, Vol. 1, No. 5 (Nov. 1948), p. 16.
72 *Decks Awash*, Vol. 7, No. 1 (Feb. 1978), pp. 29–32.
73 Interview, Harold Henriksen, 6 Nov. 1984.
74 John M. King, "An Evaluation of Canada's East Coast Sealing Industry: The 1980 Experience," BA diss., Ryerson Polytechnical Institute, Toronto, 1980 (hereafter cited as "An Evaluation of Canada's East Coast Sealing Industry), p. 89.
75 Harold Henriksen interview, 6 Nov. 1984.
76 D.L. Dunn, *Canada's East Coast Sealing Industry 1976*, p. 24.
77 Ibid., p. 22.
78 David Edwin Keir, *The Bowring Story*, pp. 383–84.
79 "From Pelts to Coats — Via the Norwegian Connection," *Gazette* (Montreal), 28 Mar. 1981, p. 24.
80 John M. King, "An Evaluation of Canada's East Coast Sealing Industry," pp. 88–89.
81 Harold Henriksen interview, 2 May 1984.

82 Cater W. Andrews, *Brief to the Special Advisory Committee to the Minister of Fisheries and Forestry on Seals* (St. John's: Government of Newfoundland and Labrador, 1971), p. 18.

83 This average share figure, taken from J.S. Colman, may be slightly askew because he inadvertently counted 17 instead of 18 vessels. *See* John Stacey Colman, "The Newfoundland Seal Fishery and the Second World War," p. 43.

84 *Trade News*, Vol. 6, No. 9 (Mar. 1954), p. 5.

85 Ibid., Vol. 10, No. 2 (Aug. 1957), p. 13.

86 *Evening Telegram* (St. John's), 6 Mar. 1954, p. 3.

87 *Trade News*, Vol. 6, No. 9 (Mar. 1954), p. 5.

89 Newfoundland. Department of Natural Resources. Fishery Research Commission, *Report No. 2* (St. John's: 1963), n.p.

90 *Evening Telegram* (St. John's), 5 Mar. 1965, p. 4.

91 Ibid., 1 Mar. 1968, p. 26.

92 Ibid., 8 Mar. 1971, p. 2.

93 D.L. Dunn, *Canada's East Coast Sealing Industry 1976*, pp. 11–13.

94 Ibid., p. 15.

95 Guy David Wright, *Sons and Seals*, p. 85.

96 Captain Morrissey Johnson interview, 6 June 1984.

Conclusion

1 British author Richard Adams has noted that the same arguments were used to defend slavery in the United States during the 19th century. *See* "The Seal Hunt: British Author Decries Slaughter," *Montreal Star*, 17 Mar. 1979, p. A8.

2 There were 4065 active sealers in 1980. John M. King, "An Evaluation of Canada's East Coast Sealing Industry," p. 40.

3 Patrick O'Flaherty, "Killing Ground," *Weekend Magazine*, Vol. 29, No. 13 (31 Mar. 1979) (hereafter cited as "Killing Ground"), p. 24.

4 John M. King, "An Evaluation of Canada's East Coast Sealing Industry," pp. 93–96.

5 Cynthia Lamson, "*Bloody Decks and a Bumper Crop*," p. 86.

6 *Evening Telegram* (St. John's), 13 Mar. 1984, p. 6.

7 Pol Chantraine, *The Living Ice: The Story of the Seals and the Men who Hunt Them in the Gulf of St. Lawrence,* trans. David Lobdell (Toronto: McClelland and Stewart, 1980), pp. 124–25.

8 Keith Thomas, *Man and the Natural World: A History of the Modern Sensibility* (New York: Pantheon Books, 1983), p. 17.

9 James Turner, *Reckoning with the Beast: Animals, Pain, and Humanity in the Victorian Mind* (Baltimore: Johns Hopkins Univ. Press, 1980) (hereafter cited as *Reckoning with the Beast*), p. 5.

10 Peter Singer, *Animal Liberation: A New Ethics* [sic] *for our Treatment of Animals* (London: Jonathan Cape, 1976) (hereafter cited as *Animal Liberation*), p. 222.

11 This is a common theme in seal-hunt protest. *See Decks Awash*, Vol. 7, No. 1 (Feb. 1978), p. 42.

12 James Turner, *Reckoning with the Beast*, pp. 124–25.

13 "The Weekend Poll: The Seal Hunt," *Weekend Magazine*, Vol. 28, No. 10 (11 Mar. 1978), p. 3.

14 Thomas Vere Philbrook, *Fisherman, Logger, Merchant, Miner: Social Change and Industrialism in Three Newfoundland Communities* (St. John's: Institute of Social and Economic Research, Memorial Univ. of Newfoundland, 1966), pp. 2–3.
15 Patrick O'Flaherty, "Killing Ground," p. 23.
16 For an early view, *see* Henry Stephens Salt, *Animals' Rights Considered in Relation to Social Progress....* (London: Macmillan, 1894).
17 Richard D. Ryder, "The Struggle Against Speciesism," in *Animals' Rights, A Symposium,* ed. David Paterson and Richard D. Ryder (London: Centaur Press, 1979) (hereafter cited as "The Struggle Against Speciesism"), p. 4. *See also* Richard D. Ryder, *Victims of Science: The Use of Animals in Research* (London: Davis-Poynter, 1975), pp. 11–26.
18 Richard D. Ryder, "The Struggle Against Speciesism," p. 14.
19 Peter Singer, *Animal Liberation,* p. 9.
20 Ibid.
21 Ibid., pp. 26–32.
22 Ibid., pp. 243–44.
23 *Globe and Mail* (Toronto), 19 Mar. 1984, p. 3.
24 Keith Thomas, *Man and the Natural World,* p. 300.
25 Ibid., p. 303.
26 Alan Herscovici, *Second Nature: The Animal-Rights Controversy* (Toronto: CBC Enterprises, 1985), p. 177.

Bibliography

Alexander, David
The Decay of Trade: An Economic History of the Newfoundland Saltfish Trade, 1935–1965. Institute of Social and Economic Research, Memorial University of Newfoundland, St. John's, 1977. Newfoundland Social and Economic Studies, No. 19.
"Development and Dependence in Newfoundland, 1880–1970." *Acadiensis,* Vol. 4, No. 1 (Autumn 1974), pp. 3–31. Fredericton, N.B.
"The Economic History of a Country and a Province." *Canadian Forum,* Vol. 53, No. 638 (Mar. 1974), pp. 12–13. Toronto.
"Newfoundland's Traditional Economy and Development to 1934." *Acadiensis,* Vol. 5, No. 2 (Spring 1976), pp. 56–78. Fredericton, N.B.

Andrews, Cater W.
Brief to the Special Advisory Committee to the Minister of Fisheries and Forestry on Seals. Government of Newfoundland and Labrador, St. John's, 1971.

Anspach, Lewis Amadeus
A History of the Island of Newfoundland: Containing a Description of the Island, the Banks, the Fisheries and Trade of Newfoundland and the Coast of Labrador. London, T. and J. Allman, 1819.

Appleton, Thomas E.
Usque Ad Mare; A History of the Canadian Coast Guard and Marine Services. Department of Transport, Ottawa, 1968.

Archibald, Samuel George
Some Account of the Seal Fishery of Newfoundland and the Mode of Preparing Seal Oil.... Murray and Gibb, Edinburgh, 1852.

Barkham, Selma
"The Basques: Filling a Gap in our History between Jacques Cartier and Champlain." *Canadian Geographic Journal*, Vol. 96, No. 1 (Feb./Mar. 1978), pp. 8–19. Ottawa.

Barr, William
"The Role of Canadian and Newfoundland Ships in the Development of the Soviet Arctic." *Newfoundland Quarterly*, Vol. 73, No. 2 (Summer 1977), pp. 19–23. St. John's.

Bartlett, Captain Robert Abram
The Log of Bob Bartlett; The True Story of Forty Years of Seafaring and Exploration. G.P. Putnam's Sons, New York, 1928.
"The Sealing Saga of Newfoundland." *National Geographic Magazine*, Vol. 56, No. 1 (July 1929), pp. 91–130. Washington, D.C.

Black, W.A.
"The Labrador Floater Codfishery." *Annals of the Association of American Geographers*, Vol. 50 (Sept. 1960), pp. 267–95. Lawrence, Kans.

Bonnycastle, Sir Richard Henry
Newfoundland in 1842: A Sequel to the "The Canadas in 1841." Henry Colburn, London, 1842. 2 vols. Vol. 2.

Brown, Cassie
Death on the Ice; The Great Newfoundland Sealing Disaster of 1914. With William Horwood. Doubleday Canada, Toronto, 1974.

Browne, Patrick William
Where the Fishers Go; The Story of Labrador. Cochrane, New York, 1909.

Busch, Briton Cooper
"The Newfoundland Sealers' Strike of 1902." *Labour/Le Travail*, No. 14 (Fall 1984), pp. 73–101. St. John's.
The War against the Seals: A History of the North American Seal Fishery. McGill-Queen's University Press, Kingston, 1985.

Byrne, Cyril
"Some Comments on the Social Circumstances of Mummering in Conception Bay and St. John's in the Nineteenth Century." *Newfoundland Quarterly*, Vol. 77, No. 4 (Winter 1981/82), pp. 3–6. St. John's.

Callanan, J.T.
"The Newfoundland Seal Hunt." In Joseph Roberts Smallwood, ed., *The Book of Newfoundland*, Newfoundland Book Publishers, St. John's, 1937–75, 6 vols., Vol. 1, pp. 69–72.

Canada. Laws, Statutes, etc.
Seal Protection Regulations. Statutory Order Regulations 64/443 (Order in Council P.C. 1964/1663).

Canadian Audubon
"New Regulations to Protect Seals." Vol. 26, No. 5 (Nov./Dec. 1964), pp. 165–66. Toronto.
"Review." Vol. 28, No. 2 (Mar./Apr. 1966), pp. 56, 70. Toronto.

Canadian Fisheries Annual
Gardenvale, Que., 1957.

Canadian Fisherman
1914–16, 1925–29, 1931. Gardenvale, Que.

Canadian Sealers' Association
Annual Report ... for 1983/84. St. John's, 1984.

Candow, James E.
"Preliminary Observations on the History of James Ryan Ltd." Manuscript on file (No. 0014[i]), Atlantic Regional Office, Canadian Parks Service, Environment Canada, Halifax, 1981.

Carroll, Michael
The Seal and Herring Fisheries of Newfoundland. Together with a Condensed History of the Island. John Lovell, Montreal, 1873.

Chambers, Edward Thomas Davis
The Fisheries of the Province of Quebec. Department of Colonization, Mines and Fisheries, Quebec, 1912. Part 1.

Chantraine, Pol
The Living Ice: The Story of the Seals and the Men who Hunt Them in the Gulf of St. Lawrence. Trans. David Lobdell. McClelland and Stewart, Toronto, 1980.

Chappell, Edward
Voyage of His Majesty's Ship Rosamond to Newfoundland and the Southern Coast of Labrador.... London, printed for J. Mawman, 1818.

Chicanot, E.L.
"New 'Eyes' for the Sealing Fleet." *Scientific American*, Vol. 138, No. 5 (May 1928), pp. 409–11. New York.

Chronicle-Herald **(Halifax)**
"Anti-Fish Campaign Has 'Bogus Premise.' " 29 Feb. 1984, p. 39.

Coish, E. Calvin
Season of the Seal: The International Storm over Canada's Seal Hunt. Breakwater, St. John's, 1979.

Colman, John Stacey
"The Newfoundland Seal Fishery and the Second World War." *Journal of Animal Ecology*, Vol. 18, No. 1 (May 1949), pp. 40–46. London.
"The Present State of the Newfoundland Seal Fishery." *Journal of Animal Ecology*, Vol. 6, No. 1 (May 1937), pp. 145–59. London.

Cotton, Sidney
Aviator Extraordinary: The Sidney Cotton Story; As Told to Ralph Barker. Chatto and Windus, London, 1969.

Davies, Brian
Savage Luxury: The Slaughter of the Baby Seals. Ryerson Press, Toronto, 1970.

Decks Awash
Vol. 7, No. 1 (Feb. 1978). St. John's.
"Shipwrecked at the Front," Vol. 13, No. 3 (May/June 1984), pp. 50–51. St. John's.

Dictionary of Newfoundland English
Ed. G.M. Story, W.J. Kirwin, and H.D.A. Widdowson. University of Toronto Press, Toronto, 1982.

Dunn, D.L.
Canada's East Coast Sealing Industry 1976: A Socio-economic Review. Department of Fisheries and the Environment, Ottawa, 1977. Fisheries and Marine Service Industry, Report No. 98.

Ellis, Frank H.
Canada's Flying Heritage. 2nd ed. University of Toronto Press, Toronto, 1980.

England, George Allan
The Greatest Hunt in the World. Tundra Books, Montreal, 1975.

Evening Telegram (St. John's)
1893, 1923, 1927, 1929–30, 1951–52, 1954–55, 1961, 1964–65, 1967–71, 1973–74, 1976–80, 1982–84.

Farnham, C.H.
"Labrador." *Harper's New Monthly Magazine*, Vol. 71, No. 425 (Oct. 1885), pp. 651–66. New York.

Farquhar, James A.
Farquhar's Luck. Petheric Press, Halifax, 1980.

Fay, Charles Ryle
Life and Labour in Newfoundland. University of Toronto Press, Toronto, 1956.

Fisher, H.D.
"Utilization of Atlantic Harp Seal Populations." In North American Wildlife Conference, Transactions of the Twentieth North American Wildlife Conference, Wildlife Management Institute, Washington, D.C., 1955, pp. 507–18.

Fisheries of Canada
"Norway Joins Canada in Ban on Hunting 'Whitecoat' Seals." Vol. 22, No. 8 (Feb./Mar. 1970), pp. 3–4. Ottawa.

Fogarty, James J.
"The Seal-Skinners' Union." In Joseph Roberts Smallwood, ed., *The Book of Newfoundland*, Newfoundland Book Publishers, St. John's, 1937–75, 6 vols., Vol. 2, p. 100.

Gardner, John
The Dory Book. International Marine Publishing, Camden, Maine, 1978.

Gazette (Montreal)
"From Pelts to Coats — via the Norwegian Connection." 28 March 1981, p. 24.

Globe and Mail (Toronto)
1979, 1982–1984.

Gosling, William Gilbert
Labrador: Its Discovery, Exploration, and Development. Alston Rivers, London, 1910.

Gosse, Edmund W., ed.
The Life of Philip Henry Gosse.... Kegan Paul, Trench, Trübner, London, 1890.

Great Britain. Newfoundland Royal Commission.
Newfoundland Royal Commission, 1933: Report. His Majesty's Stationery Office, London, 1934.

Greene, Major William Howe
The Wooden Walls Among the Ice Floes: Telling the Romance of the Newfoundland Seal Fishery. Hutchinson, London, 1933.

Grenfell,Wilfred Thomason
A Labrador Doctor: The Autobiography of Wilfred Thomason Grenfell. Hodder and Stoughton, London, [1920].

Gwyn, Richard
Smallwood, The Unlikely Revolutionary. McClelland and Stewart, Toronto, 1968.

Harvey, John
"The Newfoundland Seal Hunters." *The Canadian Magazine*, Vol. 16, No. 3 (Jan. 1901), pp. 195–206. Toronto.

Hatton, Joseph, and Moses Harvey
Newfoundland; Its History, Its Present Condition and Its Prospects in the Future. By Joseph Hatton ... and the Rev. M. Harvey.... Rev. ed. Doyle and Whittle, Boston, 1883.

Head, C. Grant
Eighteenth Century Newfoundland: A Geographer's Perspective. Mc-Clelland and Stewart, Toronto, 1976. Carleton Library Series, No. 99.

Herscovici, Alan
Second Nature: The Animal-Rights Controversy. CBC Enterprises, Toronto, 1985.

Hiller, J.K.
"The Newfoundland Seal Fishery: An Historical Introduction." *Bulletin of Canadian Studies*, Vol. 7, No. 2 (Winter 1983/1984), pp. 49–72. London, Eng.

Horwood, Harold
"Tragedy on the Whelping Ice." *Canadian Audubon*, Vol. 22, No. 2 (Mar./Apr. 1960), pp. 37–41. Toronto.

International Commission for the Northwest Atlantic Fisheries
"Report of Scientific Advisors to Panel A (Seals)." In *ICNAF Redbook*, Dartmouth, N.S., 1977, p. 96.

Jackson, Gordon
The British Whaling Trade. Adam and Charles Black, London, 1978.

Job, Robert Brown
John Job's Family: A Story of his Ancestors and Successors and their Business Connections with Newfoundland and Liverpool, 1730 to 1953. Telegram Printing, St. John's, 1953.

Johnson, Captain Morrissey
"Thirty Years at the Seal Hunt: Experiences of Morrissey Johnson." *The Livyere*, Vol. 1, Nos. 3–4 (Winter/Spring 1982), pp. 41–44. St. John's.

Jukes, Joseph Beete
Excursions in and about Newfoundland, during the Years 1839 and 1840. John Murray, London, 1842. 2 vols. Vol. 1.

Kean, Captain Abram
"Commentary on the Seal Hunt." In Joseph Roberts Smallwood, ed., *The Book of Newfoundland*, Newfoundland Book Publishers, St. John's, 1937–75, 6 vols, Vol. 1, pp. 73–75.
Old and Young Ahead, A Millionaire in Seals, being the Life History of Captain Abram Kean.... Heath Cranton, London, 1935.

"The Seal Fishery for 1934." *Newfoundland Quarterly 1901–1976: 75th Anniversary Edition.* Creative Printers and Publishers, St. John's, 1976, p. 142.

Keir, David Edwin
The Bowring Story. Bodley Head, London, 1962.

King, Clayton L.
"The *Viking*'s Last Cruise." In Joseph Roberts Smallwood, ed., *The Book of Newfoundland*, Newfoundland Book Publishers, St. John's, 1937–75, 6 vols, Vol. 1, pp. 76–85.

King, John M.
"An Evaluation of Canada's East Coast Sealing Industry: The 1980 Experience." BA dissertation, Ryerson Polytechnical Institute, Toronto, 1980.

Knight, Thomas F.
Shore and Deep Sea Fisheries of Nova Scotia. Queen's Printer, Halifax, 1867.

Lamson, Cynthia
"Bloody Decks and a Bumper Crop": The Rhetoric of Sealing Counterprotest. Institute of Social and Economic Research, Memorial University of Newfoundland, St. John's, 1979. Newfoundland Social and Economic Studies, No. 24.

Lavigne, David M.
"Counting Harp Seals with Ultra-violet Photography." *Polar Record*, Vol. 18, No. 114 (Sept. 1976), pp. 269–77. Cambridge, Eng.
"The Harp Seal Controversy Reconsidered." *Queen's Quarterly*, Vol. 85, No. 3 (Autumn 1978), pp. 377–88. Kingston, Ont.
"Life or Death for the Harp Seal." *National Geographic*, Vol. 149, No. 1 (Jan. 1976), pp. 129–42. Washington, D.C.

Lavigne, D.M., et al.
The 1977 Census of Western Atlantic Harp Seals Pagophilus groenlandicus. Department of Fisheries and the Environment/Canadian Atlantic Fisheries Scientific Advisory Committee, Dartmouth, N.S. CAFSAC Research Document, No. 77/27.

Lillie, Harry R.
The Path through Penguin City. Ernest Benn, London, 1955.

Lindsay, David Moore
A Voyage to the Arctic in the Whaler Aurora. D. Estes, Boston, 1911.

Lysaght, Averil M.
Joseph Banks in Newfoundland and Labrador, 1766: His Diary, Manuscripts and Collections. University of California Press, Los Angeles, 1971.

McGrath, Patrick Thomas
Newfoundland in 1911.... Whitehead, Morris, London, 1911.

MacKay, R.A., and S.A. Saunders
"Primary Industries." In Robert Alexander MacKay, ed., *Newfoundland; Economic, Diplomatic, and Strategic Studies*, Oxford University Press, Toronto, 1946, pp. 78–111.

Mail-Star **(Halifax)**
"Enormous Boycott Predicted." 10 Jan. 1979, p. 10.

Major, Kevin
"Terra Nova National Park 'Human History Study': A History of Southern Bonavista Bay from Alexander Bay to Goose Bay." Manuscript Report Series, No. 351, Parks Canada, Ottawa, 1979. Published as *Terra Nova National Park: Human History Study*, Parks Canada, Ottawa, 1983, Studies in Archaeology, Architecture and History.

Maxwell, Lieutenant William
"The Newfoundland Seal Fishery." *Nature*, Vol. 10 (Aug. 1874), pp. 264–67. London.

Mercer, Malcolm C.
The Seal Hunt. Department of Fisheries and Environment, Ottawa, 1976.

Montreal Star
"The Seal Hunt: British Author Decries Slaughter." 17 March 1979, p. A8.

Mosdell, H.M., ed.
Chafe's Sealing Book; A History of the Newfoundland Sealfishery from the Earliest Available Records down to and including the Voyage of 1923. Statistics Prepared by L.G. Chafe.... Trade Printers and Publishers, St. John's, 1923.

Mowat, Farley
Wake of the Great Sealers. With prints and drawings by David Blackwood. McClelland and Stewart, Toronto, 1973.

New American World: A Documentary History of North America to 1612
Ed. David B. Quinn, with Alison M. Quinn and Susan Hiller. Arno Press, New York, 1979. 4 vols. Vol. 4.

Newfoundland. Colonial Secretary's Office.
Census ... 1845, ... 1857, ... 1869, ... 1884. St. John's, 1845, 1857, 1870, 1886.

Newfoundland. Department of Natural Resources.
Report of the Newfoundland Fisheries Board and General Review of the Fisheries for the Years ... 1937 and 1938 through *Report ... 1948.* Titles and imprint varies.

Newfoundland. Department of Natural Resources. Fishery Research Commission.
Report No. 2. St. John's, 1963.

The Newfoundlander (**St. John's**)
"Outfit for the Seal Fishery at Bay Roberts." 8 Apr. 1861, p. 2.
"Outfit for the Seal Fishery at Brigus." 8 Apr. 1861, p. 2
"Outfit for the Seal Fishery, St. John's, 1861." 11 Mar. 1861, p. 3.
"Vessels Cleared for the Seal Fishery at the Port of Carbonear, for the Year 1861." 11 Mar. 1861, p. 3.
"Vessels Cleared from Harbour Grace for the Seal Fishery, 1861." 14 Mar. 1861, p. 3.

Newfoundland. General Assembly. House of Assembly.
Journal of the House of Assembly of Newfoundland ... 1891 through *... 1900, Journal ... 1914.* Imprint varies.

Newfoundland. Laws, Statutes, etc.
Acts of the General Assembly of Newfoundland. Imprint varies, St. John's, 1873–1926.
Workers' Compensation Fishing Regulations, 1980. Newfoundland Regulation 277/80.
Workers' Compensation Legislation. Statutes of Newfoundland, 1948, No. 30.

Newfoundland. Memorial University, St. John's. Centre for Newfoundland Studies.
"Reports of the Newfoundland Sealing Fleet, 1924–41," by Levi Chafe.
"Review of the Newfoundland Fisheries for 1959," by the Honourable J.T. Cheeseman, Minister of Fisheries.

Newfoundland. Royal Commission on Labrador.
Report of the Royal Commission on Labrador. St. John's, 1974. 3 vols. Vol. 3.

Newfoundland. Statistics Agency.
Historical Statistics of Newfoundland and Labrador. Division of Printing Services, Department of Public Works and Services, St. John's, 1977–. Vol. 2.

Newfoundland and Labrador. Provincial Archives.
GN1/10/0, War Correspondence.
GN2/5, Colonial Secretary's Subject Files.
P5/11/6, Job Family Papers.

Noel, S.J.R.
Politics in Newfoundland. University of Toronto Press, Toronto, 1973.

Nova Scotia. General Assembly. House of Assembly.
Journal and Proceedings of the House of Assembly ... 1828. Halifax, 1828.

O'Flaherty, Patrick
"Killing Ground." *Weekend Magazine*, Vol. 29, No. 13 (31 Mar. 1979), pp. 22–24. Toronto.

O'Neill, Paul
A Seaport Legacy: The Story of St. John's, Newfoundland. Press Porcepic, Erin, Ont., 1976. 2 vols. Vol. 2.

Parker, John
Newfoundland, 10th Province of Canada. Lincolns-Prager, London, 1950.

Pazdzior, E.
"The Fishing Industry." In Ian McAllister, ed., *Newfoundland and Labrador: The First Fifteen Years of Confederation*, Dicks, St. John's, 1966, pp. 117–32.

Percey, H.J.
Merchant Ship Construction. Brownside and Ferguson, Glasgow, 1983.

Philbrook, Thomas Vere
Fisherman, Logger, Merchant, Miner: Social Change and Industrialism in Three Newfoundland Communities. Institute of Social and Economic Research, Memorial University of Newfoundland, St. John's, 1966. Newfoundland Social and Economic Studies, No. 1.

Pimlott, Douglas H.
"The 1967 Seal Hunt." *Canadian Audubon,* Vol. 29, No. 2 (Mar./Apr. 1967), pp. 41–43, 70. Toronto.
"Seals and Sealing in the North Atlantic." *Canadian Audubon,* Vol. 28, No. 2 (Mar./Apr. 1966), pp. 33–39. Toronto.

Prowse, Daniel Woodley
A History of Newfoundland from the English, Colonial, and Foreign Records. Reprint of 1895 ed. Dicks, St. John's, 1971.

***Public Ledger and Newfoundland General Advertiser* (St. John's)**
Article, no title. 14 Feb. 1832, p. 3.

Rees, Abraham
The Cyclopaedia; or, Universal Dictionary of Arts, Sciences and Literature. Longman, Hurst, Rees, Orme and Brown, London, 1819. 39 vols. Vol. 27.

Rowsell, Harry C.
"1977 Sealing Activities by Newfoundland Landsmen and Ships on the Front: A Report to the Committee on Seals and Sealing and the Canadian Federation of Humane Societies." Ottawa, 1977. Copy on file, Centre for Newfoundland Studies, Memorial University of Newfoundland, St. John's.

Ryan, Shannon
"The Newfoundland Cod Fishery in the Nineteenth Century." MA thesis, Memorial University of Newfoundland, St. John's, 1971.
"The Seal and Labrador Cod Fisheries." Manuscript on file (No. 0014), Atlantic Regional Office, Canadian Parks Service, Environment Canada, Halifax, n.d.

Ryder, Richard Dudley
"The Struggle Against Speciesism." In *Animals' Rights, A Symposium,* ed. David Paterson and Richard D. Ryder, Centaur Press, London, 1979, pp. 3–14.

Victims of Science: The Use of Animals in Research. Davis-Poynter, London, 1975.

Sager, Eric W.
"The Merchants of Water Street and Capital Investment in Newfoundland's Traditional Economy." In Lewis R. Fischer and Eric W. Sager, eds., *The Enterprising Canadians: Entrepreneurs and Economic Development in Eastern Canada, 1820–1914...*, Memorial University of Newfoundland, St. John's, 1979, pp. 77–95.

Salt, Henry Stephens
Animals' Rights Considered in Relation to Social Progress.... Macmillan, London, 1894.

Sanger, Chesley W.
"The Evolution of Sealing and the Spread of Permanent Settlement in Northeastern Newfoundland." In John J. Mannion, ed., *The Peopling of Newfoundland: Essays in Historical Geography*, Institute of Social and Economic Research, Memorial University of Newfoundland, St. John's, 1977 (Newfoundland Social and Economic Studies, No. 8), pp. 136–51.
"The 19th Century Seal Fishery and the Influence of Scottish Whalemen." *Polar Record*, Vol. 20, No. 126 (Sept. 1980), pp. 231–51. Cambridge, Eng.
"Technological and Spatial Adaptation in the Newfoundland Seal Fishery During the Nineteenth Century." MA thesis, Memorial University of Newfoundland, St. John's, 1973.

Saunders, Gary L.
"A Trip to the Ice Recalled." *Atlantic Advocate*, Vol. 68, No. 8 (Apr. 1978), pp. 11–15. Fredericton, N.B.

Scott, John Roper
"The Function of Folklore in the Interrelationship of the Newfoundland Seal Fishery and the Home Communities of the Sealers." MA thesis, Memorial University of Newfoundland, St. John's, 1974.

Sergeant, David Ernest
"Exploitation and Conservation of Harp and Hood Seals." *Polar Record*, Vol. 12, No. 80 (May 1965), pp. 541–51. Cambridge, Eng.
"Harp Seals and the Sealing Industry." *Canadian Audubon*, Vol. 25, No. 2 (Mar./Apr. 1963), pp. 29–35. Toronto.
"History and Present Status of Populations of Harp and Hooded Seals." *Biological Conservation*, Vol. 10, No. 2 (1976), pp. 95–118. Essex, Eng.

Studies in the Sustainable Catch of Harp Seals in the Western North Atlantic. Fisheries Research Board of Canada, Arctic Biological Station, Ste-Anne-de-Bellevue, Que., 1959. Fisheries Research Board of Canada, Circular No. 4.

Singer, Peter
Animal Liberation: A New Ethics [sic] *for our Treatment of Animals.* Jonathan Cape, London, 1976.

Smith, Nicholas
Fifty-two Years at the Labrador Fishery. Arthur H. Stockwell, London, 1936.

Soederberg, Staffan, and Lennart Almkvist
"In Prospect of the Seal Hunt in Canada 1977: A Comment." Swedish Museum of Natural History, Stockholm, 1977. Copy on file, Centre for Newfoundland Studies, Memorial University of Newfoundland, St. John's.

Staveley, Michael
"Population Dynamics in Newfoundland: The Regional Patterns." In John J. Mannion, ed., *The Peopling of Newfoundland: Essays in Historical Geography*, Institute of Social and Economic Research, Memorial University of Newfoundland, St. John's, 1977 (Newfoundland Social and Economic Studies, No. 8), pp. 49–76.

Thomas, Keith
Man and the Natural World: A History of the Modern Sensibility. Pantheon Books, New York, 1983.

Tocque, Philip
Kaleidoscope Echoes, being Historical, Philosophical, Scientific, and Theological Sketches, from the Miscellaneous Writings of the Rev. Philip Tocque. Ed. Annie S. Tocque. Hunter, Rose, Toronto, 1893.

Trade News
1948–49, 1951, 1954–59, 1965. Ottawa.

Tuck, James A.
Newfoundland and Labrador Prehistory. Van Nostrand Reinhold, Toronto, 1976.

Turner, James
Reckoning with the Beast: Animals, Pain, and Humanity in the Victorian Mind. Johns Hopkins University Press, Baltimore, 1980.

Watts, G.S.
"The Impact of the War." In Robert Alexander MacKay, ed., *Newfoundland; Economic, Diplomatic, and Strategic Studies,* Oxford University Press, Toronto, 1946, pp. 219–30.

Weekend Magazine
"The Weekend Poll: The Seal Hunt." Vol. 28, No. 10 (11 Mar. 1978), p. 3. Toronto.

Wilson, William
Newfoundland and Its Missionaries. In Two Parts. To which is added a Chronological Table of all the Important Events that have occurred on the Island. By Rev. William Wilson.... Dakin and Metcalf, Cambridge, Mass., 1866.

Winsor, Cater
"Reminiscences of Rev. Cater Winsor." In Naboth Winsor, ed., *"By Their Works": A History of the Wesleyville Congregation: Methodist Church 1874–1925, United Church 1925–1974,* privately published, n.p., 1976, p. 47.

Winters, G.H.
"Production, Mortality, and Sustainable Yield of Northwest Atlantic Harp Seals (*Pagophilus groenlandicus*)." *Journal of the Fisheries Research Board of Canada,* Vol. 35, No. 9 (Sept. 1978), pp. 1249–1261. Ottawa.

Wright, Guy David
Sons and Seals: A Voyage to the Ice. Institute of Social and Economic Research, Memorial University of Newfoundland, St. John's, 1984. Newfoundland Social and Economic Studies, No. 29.

A Year Book and Almanac of Newfoundland, for 1888....
Queen's Printer, St. John's, 1888.

Index

Index